UTOPIA
AND
EXPERIMENT

ESSAYS IN THE SOCIOLOGY OF COOPERATION

---

by Henrik F. Infield

KENNIKAT PRESS
Port Washington, N. Y./London

UTOPIA AND EXPERIMENT

Manufactured by Taylor Publishing Company   Dallas, Texas

ESSAY AND GENERAL LITERATURE INDEX REPRINT SERIES

To

Edward A. Norman

Founder and President of the Group Farming Research Institute
whose friendship, mature advice, and active support
made possible the studies
on which these essays are based.

Utopia and Experiment

Essays in the Sociology of Cooperation

Table of Contents

Foreword   1

Part I

1) *Utopia and Experiment*
   Decline and Rehabilitation of Utopia. Paths in Utopia.
   Return to Utopia? The Fallacy of "Either-Or". The Ex-
   perimental Approach. Sociological Experiment. A Canon
   for Sociological Experiment. The Modern Cooperative
   Community.   9

2) *Sociology and the Modern Cooperative Community*
   The Need for a Sociology of Cooperation. "Utopian"
   Cooperative Communities. The Three Main Types of
   Cooperative Farming. The Common Features. Differ-
   ences. Theory and Practice. A Field of Social Research.   24

3) *Some Recent Developments in Cooperative Farming*
   Britain. Sweden. Cyprus. Paraguay. India. Ceylon.
   Pakistan. New Zealand.   46

4) *The Link with the Past*
   The Penn-Craft Community. Young Planners at Yellow
   Springs. Amana, The Community of True Inspiration.
   An American Folk School and the Bass Lake Farm. The
   Hutterische Gemein'. The Quest for Security.   .   .   66

5) *Saskatchewan: A Pattern of Sound Growth*
   A Spontaneous Development. The Need for Cooperative
   Farming. The Economic Needs. The Social Needs. The
   Younger Sons. The War Veterans. The Political Cli-
   mate. An Effective Demonstration.   94

6) *The Case of the Chinese Industrial Cooperatives.*   .   **125**

7) *Pains of Growth in Israel*
   The *Kvutza* Faces Change. Problems of an Internal
   Character. Problems of an External Character. Coopera-
   tive Movement and Cooperative Community. The

Second Stage of the Cooperative Movement. The Present Stage. Indicated Solutions. Vitality and Self-Examination.   142

8) *Effective Education: The Study Group and the Training Farm*
Education and Cooperation. The Antigonish Movement. Saskatchewan. The Pioneer Training Farms. Applications.   158

9) *The Urban Cooperative Community*
The Issue. Cooperation and Industry. City and Community. The City *Kibbutz Ef'al.* A Dutch Project. Marcel Barbu and the Beginning. The *Community of Work.* The Communitarian Federation. Drawbacks. A Significant Advance.   180

Part II

10) *The Sociometric Test: Its Merits and Its Limitations*
Sociometry and the "Open" Community. The Sociometric Test. Cooperative Communities are Instances of Genuine Sociological Experiment. The Merits of the Sociometric Test. Limitations and Shortcomings.   .   207

11) *Quantitative Group Comparison*
Matador and Macedonia. Matador, in Saskatchewan, Canada. Macedonia, in Georgia, U.S.A. The Studies and their Results. The Biographical Group Interview. The Sociometric Test. Theoretical Implications. The Sociological Relevance of the Findings. The Methods and their Validity. The Possibility of Cumulative Research.   224

12) *Testing of a Pioneer Training Farm*
The Zionist Pioneer Training Farms. The Aims of Research in Cooperation. The Battery. The Setting of the Study. Interpretation.   266

13) *Experimental Groups and Sociological Counselling*
The Need of Experimental Groups for Counsel. The Setting of the Study. The Tests Applied. The Personal Background. The Sociometric Group Structure. The Cooperative Potential. The Obstacles Test. Discussion of the Results. Presentation. Steps of Remedial Action.   283

14) *Summary and Outlook*
Recent Developments. *Bureau dÉtudes Cooperatives et Communautaires. (B.E.C.C.).* Other Developments.   311

Acknowledgments   320

FOREWORD

Systematic and logically consistent presentation is the goal at which the scientist aims when venturing into any field of inquiry. The less cultivated the new field, however, the more elusive will be the goal. To offer findings in a tentative, loosely organized form, may then appear pardonable, even desirable. This will be particularly so where, as in the present instance, the inquiry concerns problems of acute relevance to central issues of our times.

The area of investigation mapped out by the present collection of essays is that of the sociology of cooperation. The actual field work dates back as far as 1933. At that time the author was engaged in a sociological study of the Zionist *Kvutzot* in what then was Palestine and today is Israel. A more sustained and wider range of effort became possible only after his affiliation with the Group Farming Research Institute, established in 1941 under the name of Rural Settlement Institute. The specific task assigned to this privately financed research agency was the scientific study of cooperative farming. Upon assuming the direction of the Institute, the writer was enabled not only to collect information on cooperative farm development in all parts of the world, but also, and more importantly, to travel and visit areas of significant community growth in the U.S.A., Canada, Israel, France, and, most recently, Mexico. In the process, the need for precise methods of investigation became apparent. Consequently, a battery of sociological tests and related devices was developed and systematically applied to the different groups. The results turned out to be as stimulating to the inquiry as they proved helpful to the groups themselves.

Reports of these studies, first published in the form of

monographs and scientific papers, evoked a response sufficient
to justify the assumption that a systematic presentation of the
field and the methods of a sociology of cooperation would be
a welcome contribution to social theory in general and cooperative practice in particular. An effort in this direction has been
under consideration for some time. As in all such undertakings, however, it is neither wise nor expedient to try to hasten
matters unduly. In the meantime, the need for some basic text
material on the social aspects of cooperation appears to be
growing. This seems to be the case particularly in connection
with the increasingly important role of cooperative farming
in relation to one of the most crucial issues of our times. This
problem is, generally, how to increase food production in
underdeveloped areas. Among the possible solutions to this
problem, cooperative farming ought to be among the most
seriously considered. Attempts to apply it meet, as may be
expected, with many and often very great difficulties. These
difficulties are only in part of an economic nature. More vex-
ing, and in most cases also more intractable, is the problem
of personal relations. Important as physical facilities undoubtedly are, the success of cooperative farming depends not so
much on facilities as on the people who use them. The nature
of group formation and function is a subject which has gradually attracted the attention of the most advanced social science
today. Yet inquiry is currently concentrated mainly on so-
called "closed groups," or on groups bound by the competitive
goals of present day society. Systematic study of the factors
determining the formation and success of cooperative groups
is, however, very rare. The present essays fill, therefore, an
important gap. Exploratory as they are, they represent pertinent scientific information available on the subject. They are
intended merely as a primer to a more systematic treatment
of the subject in the future.

It is with this consideration in mind that the present essays were selected and arranged. At times, it was necessary to neglect chronology and change the original texts, to eliminate repetition; to bring information, particularly of statistical nature, up to date; to re-write several portions; and, in some cases, to combine several articles into one. All this was done with the intention of coming as close as possible to systematic exposition.

Divided in two parts—one more generally descriptive, the other more specifically analytic—the volume opens with a general discussion of the issue of science versus utopia. The introductory essay attempts to show this issue to be based on an untenable dichotomy and proceeds to define cooperative communities as what they actually are, namely genuinely sociological experiments.

The first part of the book presents general descriptions of the more significant instances of modern cooperative community development and discusses briefly their most acute problems. It reveals the great variety in communal patterns compatible with cooperative farming, ranging as it does from genuine cooperation, with its emphasis on voluntary participation and respect for the dignity of the individual, to collectivization resorting to coercion and dictate from above. It is the *Kvutzot* that emerge from this survey as the most conspicuous instance of genuine cooperation. Their experience, therefore, serves to illustrate the problems besetting even the most successful of these experiments, and the possibilities of education for cooperation suggested by the work of the so-called Antigonish Movement, the Saskatchewan Adult Education Office, and the Zionist Pioneer Training Farms. Part I of the book concludes with an essay on the French *Communities of Work* which, though in existence only since the end of the Second World War, have already made their mark by offering what appears to be a most ingenious contribution to the problem of urban cooperative communities.

Useful, even indispensable, as information gained from trained observation undoubtedly is, it constitutes only the first step towards the goal of all scientific inquiry, predictive generalization. To reach this goal, more precise instruments of research are required. How such instruments were developed and applied to the study of cooperative communities is indicated in the essays that form Part II of the volume. The chief difficulty to be overcome was the almost complete lack of tools appropriate for the study of experimental groups. The only device available was the sociometric test. Helpful as this test proved to be in uncovering the social structure of a group, it left unexplained other important aspects of the situation. Other devices consequently had to be developed, such as: the biographical group interview and its substitute, the personal questionnaire; the cooperative potential test; and the obstacles test. The three essays which form the bulk of this part present the results obtained from the application of the combination, or the "battery," of these tests to some of the communities described in Part I. The diagnostic use of the battery and its therapeutic effect upon the experimental groups themselves, finally, is illustrated by a detailed description of the study made in one of the largest of the French *Communities of Work*.

Sociological research often has to concern itself not only with things as they are, but also with things as they ought to be. This ameliorative bent is sometimes denounced as doing violence to the purity of science. For better or for worse, however, it appears to be inseparable from the exercise of sociological skills in general, and from their application to society as it presents itself today, in particular. As Lester Ward said in his *Dynamic Sociology:* "The real object of science is to benefit man. A science which fails to do this, however agreeable its study, is lifeless. Sociology, which of all science should benefit man most, is in danger of falling into the class of polite amusements, or dead sciences." [1]

That is why cooperative communities offer much attraction to fundamental sociological research. They search for the "better society" neither by slipping away into an imaginary "Nowhere" nor by relying on an equally utopian Marxist "leap from the realm of necessity into the realm of freedom." Instead, they explore, in a pragmatic fashion, the possibilities of a more satisfying mode of social and economic relations. Following the method of sociological experiment, they see as their ultimate goal the discovery of more consistent and at the same time viable norms of behavior. The latter aspect of their search should make these groups socially even more significant than their contribution to wholesome reform in agriculture and industry alone would tend to make them.

Whether they will succeed and, if successful, will be able to stem mankind's current drift into disaster, is, of course, an open question. Still, the conviction keeps growing that the peoples of the world, forced by modern technology and numerical growth into unprecedented physical proximity, must learn how to live peaceably if not cooperatively with each other, if for no other reason than for sheer survival. There are some who see in the cooperative communities islands of sanity presaging the emergence of a sounder world order. Like Saint Exupéry, they feel that in the world of today they are in fact exiles "who have not built yet their homeland." Yet, they well may find themselves to be at home wherever men and women, like those of the cooperative communities, are actively searching for ways of working and living together in peace and harmony.

---

[1] quoted in Jay Rumney and Joseph Maier, *Sociology: The Science of Society* (New York: Henry Schuman, 1953), p. 167.

PART I

UTOPIA AND EXPERIMENT

Decline and Rehabilitation of Utopia

Words, like books, have their fates, and what may be called their "social careers." Some, like the word "marshal," which originally meant a servant in charge of horses, start low and rise to high rank; others, like "knave," which to begin with meant young man, start decently enough only to end in shame. There are often real, or at least good, reasons for such rise or decline, and usually they are related to changes in social and economic conditions. A case in which political interest deliberately threw a fair name into disrepute is that of "utopia." How this happened is described pointedly in Martin Buber's book, *Paths in Utopia*.

It all had to do, as Buber demonstrates, with Karl Marx's struggle for domination in his own socialist camp. In this struggle Marx—"with Engels at his side"—based his bid for supremacy on the claim that while his own socialism was scientific, that of all the other contenders was incorrect, illusory, utopian. The "demolition" of these contenders was performed with cutting thoroughness in that section of the Communist Manifesto which deals with utopian socialism. "Utopia," a word which Sir Thomas More formed from the Greek and used to designate a fair land which existed nowhere but in his own imagination, had until then enjoyed a rather genteel reputation. It was associated with attempts to picture an ideal society with none of the imperfections of any existing one, but with all the accomplishments a given age and author could think of. Marx turned the term into "the most potent missile in the fight of Marxism against non-Marx-

ian socialism." [1] With the rise of Marxism, the power of the
missile grew. The belief in its efficacy spread beyond the ever
widening boundaries of the socialist camp until, in our own
day, "utopian" has become an epithet generally reserved for
any social, economic, or political proposition considered to
be slightly queer and thoroughly futile.

It is interesting to note that attempts to rehabilitate "utopia"
coincide with a decline of the Bolshevik Revolution in the
esteem of western intellectuals. The connection between these
two trends is obvious. Marx's scientific socialism found its
most orthodox implementation in the Soviet state. Today this
state's power rivals that of the United States. But it has failed
to make good any of Marxism's promises of liberation. The
slogans of political democracy, economic equality and, above
all, the "withering away of the state," have vanished into the
limbo of power politics which turned the "dictatorship of the
proletariat" into an apparently permanent institution. The
"better society" appears to be further from realization than
ever. Refuted by its own practice, Marxist theory has fallen
into disrepute. Utopia on the other hand, is gaining in favor.
Thus the tendency to reassess its merits is strengthened.

An indication of this trend may be found in studies like
Lewis Mumford's *The Story of Utopia* (1922); J. O. Hertzler's
The History of Utopian Thought (1926); V. F. Calverton's
Where Angels Dared to Tread (1941); Marie Louise Berneri's
Journey Through Utopia (1950); and, most recently, Negley's
and Patrick's *The Quest for Utopia* (1952). The scope and
spirit of these studies is apparent by what Hertzler states to
be the purpose of his book, namely that it "attempts to give
an unprejudiced systematic treatment of the social Utopias
as a whole. "However, the cause of utopia has still more zeal-

[1] Martin Buber, *Paths in Utopia*. Transl. by R. F. C. Hull (London:
Routledge & Kegan Paul, 1949), p. 5.

ous contemporary champions. Thus David Riesman opens an
essay on community planning with the assertion: "A revival
of the tradition of utopian thinking seems to me one of the
important intellectual tasks of today." [2] And Martin Buber
devotes his *Paths in Utopia* to the demonstration that of the
two, Marxist (or scientific) and utopian socialism, it was the
latter which "may well be clearing the way for the structure of
society-to-be (p. 14)."

Paths in Utopia

Mr. Riesman, writing in America, bases his case, pragmatic-
ally enough, on the work of an architect and his collaborator.[3]
Riesman believes that it is the architectural fraternity which
"has continued to produce and to stimulate thinking in the
utopian tradition—thinking which at its best combines respect
for material fact with ability, even enthusiasm, for transcend-
ing the given." The book he discusses, *Communitas,* deals with
problems of community planning for modern industrial society. The authors offer three models, or paradigms, of community. The basis for the first paradigm is the "American
standard of living" with its goal of efficient consumption of
luxury goods. The second, and most utopian of the three
models, is one which maps out features of a community in
which the cause of modern man's "alienation," the divorce
of production from consumption, has been overcome. The
third, finally, offers an "interim solution," a combination of
the preceding two schemes with emphasis on immediate realization. What Mr. Riesman finds valuable in the contribution
of the Goodmans "lies not in this or that detail but in their

[2] David Riesman, "Some Observations on Community Plans and
Utopia," The Yale Law Journal (December 1947), pp. 173, ff.
[3] Percival and Paul Goodman, *Communitas: Means of Livelihood
and Ways of Life* (Chicago: University of Chicago Press, 1947).

explicit attachment to the now languishing tradition of utopian thought."

The material Martin Buber uses to state his case is of a more philosophical nature. He attempts "to sketch the picture of an idea in process of development." To this end he employs a method not easily followed by one not versed in Buber's particular brand of philosophy. To interpret it would require a full exegesis. Space, however, does not permit more than a brief restatement, as far as possible in Buber's own terms, of the points most relevant to our discussion.

Buber defines his concepts by means of setting them off against their opposites, and proceeds to demonstrate their validity by applying them successively to contrasting systems of thought. The first antithesis is that of "Utopia" and "Revelation." Both, he finds, originate from the same need of "the longing for that 'rightness'" or vision of "what should be" which "cannot be realized in the individual, but only in human society." While revelation, as "messianic eschatology," is "realized in the picture of perfect time," utopia is "realized in the picture of perfect space." Both aim at perfection; but while for revelation and eschatology the decisive act of cosmic perfection happens "from above," for utopia, the perfect society it aims at, is brought about by deliberate human action. Though both have "the character of realism," the realism of eschatology is "prophetic," and that of utopia "philosophical."

The conceptual juxtaposition that is central to the theme of the book, however, is that between utopian socialism on the one hand, and scientific socialism, or Marxism, on the other. Reduced to its simplest terms, the contrast can be stated as follows: scientific socialism is "necessitarian" and "apocalyptic"—utopian socialism is "voluntaristic" and "prophetic." The former places the arrival of the perfect state in the period after the final revolution, when the state will "wither away" and humanity will "leap out of the realm of necessity into the realm of freedom"; the latter holds that if the perfect state is ever to come, its coming must be prepared by immediate attack upon and remedy of the inadaquacies of society. The means used by Marxism to achieve its goal of final release from "necessity" are compulsion, centralization, regimentation; utopian socialism by contrast insists upon the fullest possible freedom of the individual, voluntary association and communal autonomy. Regarding ends, Marxism is just as "utopian" as utopian socialism. Both agree on "what should be," freedom. They decidedly disagree, however, on the means to be used to attain this end. Utopianism's choice appears to be both the more moral and expedient. It seems more reasonable to seek liberty by means of progressive liberation; it is hard to see how it can result from continued coercion.

Having thus clarified his concepts, Buber traces them through a series of contrasting systems of thought. The systems dealt with are on the one hand those of Marx, Lenin, and Stalin, and on the other those of Proudhon, Kropotkin, Landauer, as well as St. Simon, Fourier, and Owen. Buber's analysis is at its best here. Essentially, the contrast between these systems, according to him, can be stated as follows: While the system of the scientific socialists, of Marx and his followers, is based primarily on political action that of their antagonists, the "utopian" socialists, Proudhon, Kropotkin, Landauer rests primarily on social action. Marxism sets all its hope in revolution, utopianism in reconstruction of society. The former considers a centralized state as indispensable, the latter insists on decentralization and the autonomous community. The total state becomes the main instrument of the one, a federation of independent, self-governing social units, a "communitas communitatum," the chief means of the other.

Of particular interest to us are the data concerning the relation of each to the cooperative movement. The consumers', producers', and what Buber calls the "full" cooperatives[4] are associations for economic self-help, mainly. They have implemented experimentally many of the ideas of utopian socialism. Marx, the early social democrats, and later Lenin and Stalin, continuously wavered in their appraisal of these experiments. The ambivalence of their attitude ranges all the way from Marx's early dismissal of "the little experiments, inevitably abortive," to Stalin's acclaim and acceptance of the "full" cooperative, the Kolkhoz, as a welcome instrument of agricultural socialization. In all this wavering, however, one thing remained constant: whenever political expediency indicated a possible usefulness of the cooperatives, the only way they would be accepted was as weapons in the political struggle. Utopian socialists, on the other hand, always saw in them potential seeds of a new society and tried to cultivate them for their own sake just as much as for the sake of larger sociopolitical units.

As a refugee from Germany and professor of social philosophy at the Hebrew University in Jerusalem, Martin Buber became well acquainted with the *Kvutza.* The immediate knowledge of the *Kvutza,* he says, was the occasion of writing *Paths in Utopia.* To him, the *Kvutza* demonstrates that the assumption of utopian socialism is correct. It is the "experiment that did not fail."

Return to Utopia?

However much Buber's and Riesman's studies differ from one another in method, scope and subject matter, they are based on the same premise. The premise is the contrast between the utopian and the scientific way of thinking. In Riesman's essay the protagonists are utopian thinking versus com-

---

[4] What Buber calls "full" cooperatives *(Vollgenossenschaften)* has been termed "integral" cooperatives by Henri Lasserre and *"comprehensive"* cooperatives by this author.

mon sense; in Buber's book they are utopian versus scientific socialism. Riesman, in a limited, and Buber in a more elaborate way, try to show how under the impact of Marxist theory utopianism in every form became discredited. Both consider such defamation as hardly justified, particularly when one can show, as Buber does, that Marxism itself is not free from "hidden" utopianism. All the same, Marxism played a significant part in the European proletariat's struggle for political power, while utopian socialism sulked in obscurity. The establishment of the Soviet State made a demonstration of Marxism's merits possible in practice. Many socialists outside Russia put all their hopes upon this event. When the demonstration failed to come off, their hopes turned into bitter disillusionment. The issue between utopian and scientific socialism was now ready for a new turn. The argument is reversed. Since Marxism proved itself wrong, therefore, utopian socialism must be right. On this point both authors are in complete agreement. Consequently, the solution they propose is essentially the same: return to utopia.

At first glance, the argument seems plausible enough. Upon closer examination it loses much of its persuasiveness. It is based on assumptions that are by no means beyond questioning. Is it true, for example, that the achievement of a "better society" is possible only by means of either utopian or scientific socialism? Are these really the only alternatives we may think of? What about Henry George and Edward A. Filene, the "single tax" or "intelligent" mass production as a basis for "successful living in this machine age?" As for the "better society," by which criteria do we judge better or worse? Good or better for whom, and under what conditions?

The Fallacy of "Either-Or"

The main weakness of the argument is its "either-or" character. Recent social science research has taught us that such dichotomies, plausible as they appear at first sight, have little relevance to social facts. What the social scientist encounters in social reality are not absolute entities in logical juxtaposition, but continuous variables of one or another kind of behavior. Buber seems to be aware of the scientist's approach to the matter when he approvingly quotes Gustav Landauer's definition of the state: "The State is a condition, a certain relationship between human beings, a mode of human behaviour; we destroy it by contracting other relationships, by behaving differently (p. 46)." So far as his own argument goes, however, Buber largely ignores Landauer's anticipation of the approach of modern social psychology and sociology. For State and society are no longer regarded as objective entities, but as modes of human interaction, occurring under certain social conditions and in given cultural contexts. From the point of view of pragmatic social science, we must in fact question the usefulness of such generalized concepts as "genuine human society," and the validity of sweeping claims such as "Common management is only possible as socialistic management," or "The primary aspiration of all history is a genuine community of human beings—genuine because it is community all through (p. 133)." These generalizations no doubt express some noble sentiments, but what can the critical investigator of societal phenomena profit from them?

The alternative between utopian and scientific socialism does not appear to be very helpful. Buber himself, as we have seen, points up the "hidden" utopian traits in scientific socialism. He could just as easily have shown the scientific traits in utopianism. Under these circumstances, it makes little sense to treat science and utopia as mutually exclusive propositions. In some respects they differ and in others they are alike. We ought to find out to what extent they do and to what they don't. In this way we would at least arrive at some tangible facts. The issue will then not have to remain locked within a sterile "either-or" where disillusionment with science is left the sole alternative of a return to utopia. Instead, it might lead to the more fruitful question of how much of each can be used to improve the working of our society.

The term "utopian" has been abused to a point where the value of its further use has become debatable. Its continued application to experimental communities is merely confusing. Why classify as "utopian" a community like the Hutterite which has existed now for more than four hundred years and is still growing? Because it differs in its ways of life from those practiced outside it? But at a time when change is recognized as the main characteristic of all social existence, it makes little sense to call deviation from a rule of behavior by a eulogistic or derogatory name. If any change is "utopian," then all changes must be so considered, and the term loses its distinct meaning. Discussions operating with the concept "utopian" hardly can be considered as pertinent, and any proposition derived from such discussion, including that of a return to utopia, becomes, to say the least, empty.

The Experimental Approach

We do not mean to infer that discussion of the issues in question is in itself futile. This is no mere squabble of terminology. To make sense within the framework of modern social science the inquiry must be conducted on a more factual level. The issue then ceases to be one between ideological systems. It reduces itself to the more concrete question of why people behave in a way denoted by the term "utopian."

The motive power in all human activity is need. The need for a different or "utopian" kind of behavior arises when dissatisfaction with a given social situation activates in the people concerned the desire not merely for partial changes—or "reforms"—but for a total change of the situation by non-violent means. Behavior of this kind may be confined to the imaginary level alone. It may also extend beyond it into practical enactment. In the first case, the result would be "utopia" as literature. In the second, we would be witness to the attempts at realization of creeds, socio-reformistic blue-prints, etc., as exemplified by the various "utopian" communities. Both, the imaginary as well as the active attempts at total social change are in the nature of sociological experiments. The blue-prints may be considered as mental models of such experiments, and the communities as actual experiments.

This distinction between mental models and actual experiments makes possible a more accurate description of the essentially similar character and identical purpose of the varied and often conflicting attempts at social reconstruction. The differences between these attempts will then become a matter of degree rather than kind. We require a basis for critical evaluation, and scientific comparison. Evaluation and comparison would have to be in each case in terms of the felicity and fertility of the basic premise, or hypothesis, from which the experiment proceeded; the kind and degree of external and internal resistance it met; the implicit or explicit methods used in controlling each phase of the experiment; and, finally, the degree of precision obtained in the validation of results. A quantitative approach of this sort can in principle be applied to the most short-lived tiny community as well as to such a giant undertaking as the collective agriculture of the Soviet Union. More importantly, the employment of objective methods of investigation permits verification, control, and prediction. The investigation would develop criteria in terms of which the study of past and present experiments in cooperative living would yield cumulative results. These, in turn, could well serve to raise the standards of future experiments in the same direction.

Sociological Experiment

At this point it may be well to stop and ask more specifically: what is meant by "sociological experiment?"

Attempts to clarify the issue of experiment in sociology have by no means reached the stage beyond controversy. The social scientist who holds the methods of the natural sciences to be the only valid ones, faces a real dilemma with regard to sociological experiment. Either his ethical scruples compel him to reject a scheme proposing to handle human beings as if they were mere physical objects or animals; or he simply makes the most of data which do not relate immediately to sociological subject matter, the manifest and potential interaction of human beings in a given social and cultural context. In the first case, he rejects experiment altogether. In the second, his procedures will tend to remain purely statistical. "Experiments" of this kind—such as the much talked about "ex-post-facto" experiment[5]—have little bearing on human groups in interaction. Statistical exercises of this kind may have some marginal relevance to sociology, but they are not sociological experiments.

However, the question is not one of either natural science experiment, or no experiment at all. It is rather: Can there be an experiment which, to use Poincaré's terms, "accords" with the "properties" of societal facts? If so, what method would such an experiment have to follow? To answer the first part of the question in the affirmative, we need only modify Henri Poincaré's famous dictum about geometry and its convenience.[6] One experimental canon, we might say, cannot be

[5] See: Ernest Greenwood, *Experimental Sociology* (New York, King's Crown Press, 1945), particularly chs. IV and VIII; and Stuart F. Chapin, *Experimental Design in Sociological Research* New York, Harper & Brothers, 1947), *passim*. See also J. L. Moreno, "Sociometry and Marxism," *Sociometry,* XII, No. 1-3 (Feb.-Aug. 1949).

[6] Henri Poincaré, *Science and Hypothesis* (New York, The Science Press, 1905), p. 39.

more true than another: it can only be more convenient. Even John Stuart Mill's classical canon of experiment, based on the logical categories of "agreement" and "difference," accorded with the "properties" of physics only so long as scientific scrutiny remained relatively unrefined. With the development of more refined instruments of scientific analysis, interpretation in terms of "one track" cause and effect relationships gave way to one based on the concept of the dynamic field in which cause and effect can be assigned to one and the same factor at the same time.

A Canon for Sociological Experiment

Where, as in sociology, the subject matter by definition concerns interaction, the field, or situation, is a concept basic to all analysis. To lift particular cause and effect relations of neatly isolated factors from a social situation is not actually feasible.[7] If it were, it would not add anything significant to our understanding. A method in accord with the essential "properties" of the social situation must seek to do justice to the basic mechanisms of social behavior—even at the risk of neglecting some of the most strictly logical "properties." A sociological experiment, we may formulate tentatively, starts with people who in a given situation find themselves barred from the satisfaction of some pressing needs and deliberately proceed to create for themselves a new and contrasting situation which they expect to yield the desired satisfactions. Activated by the common intent, each member of the group is ready to experiment with himself as well as with the possibilities of the new situation. The participants "control" their self-induced experiment by checking each step in the initiated

[7] For a most recent statement on such valiant but unrewarding effort, see Samuel A. Stouffer, "Some Observations on Study Design" *American Journal of Sociology,* LV, 4 (January 1950), pp.

change against their own sense of achieved satisfaction. They "validate" it by the extent to which they agree among themselves about each phase of its outcome. If agreement is unanimously in the negative, the given phase of the experiment is abandoned. If there is no unanimity, the dissenting members may cease to participate in the experiment, while the others consolidate the effected changes.

Such a canon may be more complex and less pointedly logical than that of the natural sciences. It however has the decided advantage of deriving from a method which marks out its steps in terms of manifest and potential human interaction in a given social context. Being a general rule, this canon represents a state of perfection which is rarely, if ever, attained in actual conduct. The experimental communities and other related social experiments at best only approximate it. It may be said, though, that the success or failure of any given experiment depends upon the degree to which it deviates from this canon. The instance of modern cooperative community generally considered as the most successful—the Israeli *Kvutza*—comes closest to our model. The *Kvutza,* as we have shown elsewhere,[8] was initiated spontaneously by people who in a given situation found themselves unable to satisfy some of their basic needs. By deliberate action they drastically modified the basis of their economic relations. They pooled their resources and decided to own and operate everything in common. This drastic modification necessitated changes in the social structure. Step by step such changes were introduced with the consent and under the control of all, and were institutionalized when found validated in terms of the satisfactions they yielded. In this way there came into existence a type of community characterized by an extreme degree of compre-

[8] See H. F. Infield, *Cooperative Living in Palestine* (New York, Henry Koosis & Co., 1948).

hensive cooperation. It became of central significance in the Jewish colonization of Palestine and provided important impulses for the establishment of the State of Israel.

The Modern Cooperative Community

In dealing with the *Kvutza,* Martin Buber touches upon this important aspect of "an experiment that did not fail." In one place (p. 146) he speaks of the Jewish communal settlements as of "an experimental station where, on common soil, different colonies or 'cultures' are tested out according to different methods for a common purpose." At once, however, he returns to an interpretation of the *Kvutza* in terms of the two poles of socialism and a dramatic "either-or." The fateful choice before us today is between the one pole which "we must designate . . . by the formidable name of 'Moscow' "; and the other which Buber "would make bold to call 'Jerusalem,' " the cooperatively structured commonwealth of Israel with the *Kvutza* as its basic unit, or as he prefers to call it, its "cell."

"Jerusalem," which by implication also stands for the final success of utopia, is obviously Buber's own choice. As a social philosopher he has a right to state, and solve, his problems in terms of absolutes. Even though we agree with Riesman and Buber that a re-examination of utopian thinking is in order, we can not be satisfied by simply reverting to it. There is much to be learned from the cooperative communities, and particularly the *Kvutza,* but hardly by classifying them as "utopian." These communities are rather of a kind with those early work-a-day attempts at cooperation which made possible the first successful consumers' store. Hundreds of "Union Shops" failed before the Rochdale Weavers, learning from these failures, established their store. Hundreds of "utopian" communities came to naught before the first modern cooperative community, the *Kvutza* succeeded. All of these communities are parts of the cooperative movement. Like the movement itself, they are the outgrowth of a situation in which people, finding themselves unable to satisfy some basic needs, resorted to cooperation because it promised to yield the desired satisfactions. The tremendous growth of cooperative associations all over the world—from the one consumers' store at Rochdale to a world-wide movement which at the beginning of World War II counted over 800,000 societies with some 150 million members in fifty-seven civilized countries—should serve as proof that the promise was by no means unrealistic. In this development, the communities represent merely the most advanced stage.

To us, the paramount significance of the cooperative communities lies in what we have called their experimental character. Most social analysts today seem to agree that the crisis we are experiencing stems from the fact that ours is a period of far-reaching changes. These changes, Toynbee and Sorokin believe, indicate that we find ourselves in a process of transition to a new phase of civilization. If this is true, the future of humanity depends on the direction which this transition will take. It would, at any rate, be of great advantage to be able to control it, not by hopes and illusions, but by scientific insight based on pertinent and observable facts. Such may be gained today above all in the socio-economic laboratories which are the cooperative communities.

Philosophical rehabilitation of utopia cannot, it seems to us, be of much help in the development of effective instruments of attack upon the "crisis of our times." Systematic observation and scientific interpretation can. What we need are precise methods and techniques of investigation, and personnel trained in their use. The social scientist's task is undoubtedly more time-consuming, exacting, and unrewarding in terms of easy generalizations; but, by the same token, his contribution to the solution of some of the main problems of our time may,— so at least he hopes—be more solid.

2

SOCIOLOGY AND THE
MODERN COOPERATIVE COMMUNITY

The Need for a Sociology of Cooperation

In his last and undelivered address, Franklin D. Roosevelt summed up the most urgent need of these our troubled times in this sentence: "We are faced with the pre-eminent fact that if civilization is to survive we must cultivate the science of human relationships—the ability of people of all kinds to live together and work together in the same world, at peace." The inauguration of the atomic era, frightfully manifested by the dropping of the atomic bomb on Hiroshima, has lent to these words a crucial poignancy.

The science which considers human relationships as its specific area of investigation is sociology. The mode of human interrelations based on doing things together is cooperation. From which it follows that, if Roosevelt was correct, the fate of our civilization will be determined by our ability to cultivate the sociology of cooperation. Happy in his choice of words to the very last, Roosevelt deliberately used the term "cultivate." For although cooperation appears to attract more and more the attention of social scientists, our knowledge of it is as yet "scattered, spotty, and even chaotic." This applies not only to methods, but even to the bare facts of cooperative development. How otherwise could the significant and relatively extensive growth of the cooperative community in the last decades pass, as far as sociology is concerned, practically unnoticed? In the Soviet Union the co-operative settlement, the *Kolkhoz*, has almost completely replaced all other forms of rural organization; several thousand collective *Ejidos* at one time were established in Mexico; in Palestine unique feats of resettlement have been performed by agricultural communities practicing a most comprehensive type of cooperation, the *Kvutzot*. Yet hardly a reflection of these facts can be found in sociological literature.[1]

A short survey of the development and functioning of the modern cooperative community may serve, therefore, a useful purpose. It may help direct attention to a rich store of cooperative practice particularly suited to investigation of the kind which may lead to the foundation of a sociology of cooperation.

According to their origin, we can distinguish three basically different types of cooperative communities: (1) the religious; (2) the socio-reformistic; and (3) those predominantly motivated by economic considerations.

The religious and the socio-reformistic communities, frequently characterized as "utopian," were predominant in the past. They made their appearance at different times since the early 16th century, but most of them were in existence during the 19th century in the New World. Of 130 "utopian" community enterprises, representing 236 community units, which are known to have existed in the United States of America and in Canada, only the religious have survived for a longer period. Several of these lasted for more than a century; three, the Amana Community (founded in 1714), the Doukhobors (organized about the middle of the 18th century), and Hutterite colonies (first formed in 1528), are still in existence.[2]

The most significant of the "utopian" communities are the Hutterite colonies. According to a census, taken in 1951 (published in the Mennonnite Quarterly Review of January, 1951) Hutterite colonies are found today in addition to South Dak-

[1] To fill this gap the present writer presented available information on the subject in his *Cooperative Communities at Work* (New York: Dryden Press, 1945; London: Kegan Paul, 1947).

[2] See Joseph W. Eaton and Saul M. Katz, *Research Guide on Cooperative Group Farming* (New York, H. W. Wilson, 1942).

ota, in North Dakota and in Montana, in the United States, and in the Canadian provinces of Alberta and Manitoba. There are also some newer colonies in Ontario, Canada, in England and in Paraguay. The total number of all these colonies at present is 96, and their population is 9,211. Virtually all this growth is due to natural increase of the Hutterite population. Only five of the colonies, counting 711 souls, consist of converts.

The census distinguishes three kinds of colonies. It calls Kinship Colonies those whose members are almost exclusively descended from the original Hutterite immigrants; only 38 of the members stem from Mennonites, who themselves are culturally and genetically closely related to the Hutterites. Each of the Kinship Colonies is affiliated with one of the three federative groups of the *Schmiedenleut, Dariusleut,* or *Lehrerleut.* Unaffiliated Kinship Colonies are those, although of Hutterite descent, rejecting the authority of the Hutterite church or not recognized by it. There are altogether four such colonies, all founded between 1924 and 1944, and all located in Alberta. Their population totals 101. Finally, there are the Convert Colonies, formed as the name indicates, by people of different backgrounds who accept the Hutterite faith and a communal way of life. There is one such colony in Canada in Ontario; it consists of 46 persons of partly Hungarian and partly American-Canadian descent. The *Bruderhof* in England, with 165 people, and the three *Bruderhoefe* in Paraguay, with an estimated population of 700, were founded by converts of partly German, partly English and other West European origin.

The Hutterite community is typical of all the "utopian" communities founded on religious principles and practicing all-inclusive or comprehensive cooperation. It offers all the advantages derived from doing things together, such as economic security, a high degree of work-satisfaction with the ensuing mental and emotional health, and a strongly developed we-feeling which makes for social security. Frugality and isolation, fostered for the sake of purity and preservation of their creed, enable them, however, to attract only small numbers of participants.[3]

No instance of equal longevity can be found among socio-reformistic communities. The longest-lived single community of this type, the Llano-Cooperative Colony, lasted only 22 years.

One aspect of the "utopian" community may be worth noting here. It is true that in most instances its life span was very short. But contrary to popular opinion, failure was rarely due to economic causes, even in the case of the socio-reformistic type. The North American Phalanx, a Fourierist group, paid dividends of 5 or 6 per cent; and the Icarian colonies, as Gide relates, despite extraordinary hardships, "did not die of poverty. They carried on somehow or other, and some of them even finished up in comparative comfort." Dissolution in most cases came about, rather, because of lack of experience in agriculture, and because of inability of members to get along with each other.

The Three Main Types of Cooperative Farming

While the "utopian" communities were motivated by religious creeds or socio-reformistic zeal, the modern cooperative community—in its main types synonymous with the cooperative farm—has developed into a new socio-economic kind of organization, used by governmental or semi-governmental agencies as an instrument of rural rehabilitation. The growth of interest in the practical possibilities of cooperative farming has been very marked, particularly since the end of World War II, in many countries of Asia and the Far East. The soundness

[3] For further details see this author's *Cooperative Communities at Work,* Chapter II.

of this trend was affirmed at a meeting of the Food and Agriculture Organization of the United Nations held in Lucknow,
India, from October 24th to November 2nd, 1949. At its
conclusion, this meeting—attended by 38 delegates, observers,
and representatives of the FAO from Australia, Ceylon,
France, India, Pakistan, Thailand, and the United Kingdom—
resolved, among other things, that "Cooperative farming provides an ideal solution for the pooling of resources of the
cultivators in land, labor, and capital." Accordingly, the meeting recommended "that Governments should actively encourage cooperative farming societies with financial and technical
advice." [4]

It might well be expected that this significant resolution of
the FAO will have a stimulating effect upon attempts at cooperative farming, possibly even beyond the areas directly
represented at the meeting. Studies like *Co-operative Farming*
(Bombay, 1949), published by the Agricultural Credit Department of the Reserve Bank of India, indicate that this expectation is not in vain. The instances therein described of experiments already being undertaken in parts of India, are a sign
that the issue has ceased to be one of deliberation and has
rather become one of effective implementation.

Cooperative farming is no longer an altogether novel device.
It has been used for a number of years in several countries.
In particular, the aim may be the total reorganization of agriculture, as in the Soviet Union; the rehabilitation of the low-
income farmer, as in Mexico; or the solution of certain crucial
problems of resettlement, as in Palestine or what is now Israel.
In all these instances the primary motive is predominantly
economic. Thus, although the modern cooperative community
resembles that of the past in many ways, it differs from it
essentially in origin and in basic objectives. A brief comparison

[4] Report of Technical Meeting on Co-operatives in Asia and the
the Far East. FAO of the United Nations, December, 1949.

of the essential features of the three main types of cooperative farms existing today, the *Kolkhoz,* the collective *Ejido* and the *Kvutza* or *Kibbutz* should help to illustrate the point.

The Common Features

There are several basic features which are common to the three main instances of the modern cooperative farm; there are others in which they differ. To begin with those they have in common, we turn first to the motive of origin. It is essentially the same in all three instances. All three have their origin in practical, socio-economic needs directly related to problems of agriculture.

In Soviet Russia the specific need arose from conditions in which agriculture found itself as a result of the Bolshevik Revolution, carried out under the slogan, "Peace and Bread." Once victory was assured, all land was distributed among the peasants. The result was a splitting up of all the land into some 25,000,000 small farms, each of them capable of producing barely more than was needed by the peasant's own family. Little, if anything, was left to supply the cities. To run his farm, the small peasant needed credits, and obtained them from the wealthier farmer, the *kulak.* Both the deficiency of marketable output and the dominance of the middle-class *kulak* presented to the new Soviet state grave problems which had to be solved if the rural economy was to be brought in line with the intended development of the country.

To raise the marketable output, farming had to become mechanized. This required, first, increase of industrial output, and second, the most efficient distribution and utilization of available machinery. Both requirements could best be fulfilled if the peasants could be induced to pool their land and to use the machinery in common. Large-scale cooperative farming would solve not only the problems of marketable output, but would also break the dominance of the *kulak.* In addition, it would reduce the number of hands in agriculture, and thus free them for use in industry, the expansion of which was, in turn, the *sine qua non* of the mechanization of agriculture.

It was thus that, at the Fifteenth Party Congress, collectivization was decided upon, and that, under the first Five Year Plan (1928-33), energetic attempts were made to induce the peasants to form *artels* of the type known under the name of *Kolkhoz*. In selecting the *artel* type of organization in preference to the more extreme *commune* and the looser *Toz*, the main consideration was that of expediency. As Stalin himself puts it, the *artel* was deemed to be a "simple affair," and more acceptable to the broad mass of peasants.[5]

Concern with similar problems of Mexican agriculture marked the intensive fostering of the collective *Ejido*. These problems were referred to by Lázaro Cárdenas, in the decree initiating the establishment of such *Ejidos* in the Laguna district, as the two responsibilities of the *Ejido:* (1) . . . as a social system it must free the peasant from the exploitation to which he was subject under the feudal and the individualistic regimes; and (2) . . . as a mode of agricultural production, it must yield enough to furnish the nation with its food requirements.[6]

The motives behind the origin of the *Kvutza* were, similarly, related to problems of agriculture. Here, a particularly acute situation arose in connection with the requirements of Zionist resettlement. The development of Jewish agriculture in Palestine faced two main obstacles: (1) the extremely poor quality of available soil; and (2) the almost complete lack of agricultural experience on the part of the prospective settlers. Progress along the lines of traditional individual settlement proved to be so slow as to make prospects for success in a near future

[5] See Joseph Stalin, *Building Collective Farms* (New York, 1931), p. 106.

[8] Cf. Nathaniel and Sylvia Weyl, *The Reconquest of Mexico—The Years of Lázaro Cárdenas* (New York, Oxford University Press, 1939), p. 220.

very doubtful. The only alternative which offered itself under these circumstances was that of group-settlement. There was, as Arthur Ruppin, the head of the Palestine Office, later put it, little choice in the matter. The question appeared to be one of either group-settlement or no settlement at all. The type of settlement that emerged has since become widely known under the name *Kvutza* or *Kibbutz*.

Virtually no difference exists between the *Kolkhoz*, the collective *Ejido*, and the *Kvutza* as to their theoretical adherence to the principles of cooperation. The internal administration of all three is based on the Rochdale Principles. It is only that, true to their nature as communities, all three had to modify some of these principles to make them fit their specific requirements.[7] One of the principles is that of open membership. Community implies more than some limited economic activity; it means living as well as working together. Moreover, community is also naturally restricted by the extent of the geographic area on which it is located. Because of these and other reasons membership in a community cannot be open in the same sense as it is, for instance, in a consumers' store. For this reason the admission of members is subject to requirements stricter than those imposed in cooperatives of more limited aims. None of the communities in question allows, however, any restriction because of race or religion.

Another principle which had to be modified when applied to the concrete community situation is that of distribution of dividends according to the amount of purchase. Since the most important aspect of participation in a cooperative farm is that of shared labor, distribution of net profits according to the amount of purchase would make little sense. The practice followed in all three instances is, rather, to take the amount of

[7] For a pertinent discussion of this problem, see Morris Mitchell, "Extension of Rochdale Principles to Co-operative Community," in: Henrik F. Infield and Joseph Maier (Eds.), *Co-operative Group Living* (New York, Henry Koosis & Co., 1950), p. 144 ff.

labor contributed as the main basis for the equitable distribution of profit.

As to the remaining principles, the practice, in all three instances, is identical with that in any other genuinely cooperative association. No member has more than one vote; only limited interest, if any, is paid on investment; members have equal rights and privileges, independent of sex; there are regular meetings at which the members participate in decision-making; and finally, rules of proper auditing are observed.

All three types of community are further alike as to their basic economic activity, which is agricultural production carried on cooperatively.

Another point on which there is, at least in theory, no difference between *Kolkhoz,* collective *Ejido,* and the *Kvutza* is that of their internal autonomy. In all three, the General Assembly of all members is designated as the highest authority in all internal affairs of the group. In all three there prevails also the practice of delegating the conduct and supervision of the community's business to committees elected by the membership for a limited length of time. Admission, punishment, and expulsion of members rest, by law, in the hands of the General Assembly.

Finally, in all three instances there can be observed certain changes in the social institutions in general, and in the character of the participants in particular. In all types of cooperative farms, family, education, recreation, care of the sick, invalid, and aged are, to a minor or major degree, affected by the spirit of cooperation. The position of woman becomes more nearly equal to that of man. Education becomes more group-centered and extends beyond formal schooling to work participation on the part of the children, and to more or less directed self-education on the part of the adults. The sense of mutuality tends to increase and makes care of the sick, the invalid, and the aged a matter of common concern. Although objective methods of observation, comparison, and evaluation applicable to changes in psychological traits have yet to be developed, most observers are in agreement as to the effect of cooperative living, in all three instances, upon the behavior of those who have been participating in it for some length of time. There appears to take place a certain intensification of all modes of cooperative behavior enhancing the ability of people of all kinds to get along with each other.

Differences

The fact that *Kolkhoz,* collective *Ejido,* and *Kvutza* are alike in certain basic features makes it possible to group them together as generically of one kind. This, however, does not mean that there are not considerable differences between them. Such differences are due mainly to the varying cultural and political backgrounds under which the cooperative farms exist. Most marked, as we shall see, is the contrast between the *Kolkhoz,* on the one hand, and the *Kvutza* on the other. In many respects the collective *Ejido* can be considered as a subtype of the *Kolkhoz.*

Although all three types of cooperative farms resemble one another in the motives of origin, they differ widely in the manner of organization. The *Kolkhoz* and the collective *Ejido* owe their establishment to administrative measures. The *Kvutza,* on the other hand, grew, step by step, out of the spontaneous decisions of those who first shaped its basic socio-economic structure. This difference is most marked in the respective membership requirements of the three types.

A *Kolkhoz* is formed when several peasants living in the same neighborhood decide—or are induced to decide—to socialize their *basic means of production,* meaning labor, soil, draft beasts, farm structures, and implements, while keeping their individual homes, small gardens, a bit of livestock, poultry and the like, for themselves. Membership is open to all toilers who have reached the age of sixteen, and who are willing to comply with the established rules and regulations. Application for membership to an already established *Kolkhoz* is taken up, first, by the Management Committee of the *Kolkhoz*, and is, legally, subject to the approval of the General Assembly. Excluded from membership are *kulaks* and people deprived of their civil rights. Exceptions are made in the case of families who count among their members soldiers, sailors, or village teachers willing to recommend the applicant. An interesting sidelight on the effect of collectivization when ordered from above, is the provision barring peasants "who before joining the collective farm, slaughter or sell their cattle get rid of their stock, or wantonly sell their seed corn." [8]

Only slightly different in nature are the requirements for the formation of a collective *Ejido*. There must be at least twenty eligible male peasants forming a group to petition the Government for land. To be eligible, these peasants must: (1) be Mexican by birth; (2) over sixteen years of age; (3) for at least six months preceding group formation resident in the area in which the *Ejido* is to be formed; (4) agriculturists by profession; and (5) own no more than 2,500 pesos, or be of similar low income status. If the group can lay claim to land that once belonged to them, the land is *restored* to them; if their only claim is landlessness, land expropriated from wealthy landowners *(haciendados)* is *donated* to them. Both processes are quite protracted and cumbersome, and open to many profiteering practices by the peasants' own attorneys as well as the administrative personnel. The allotted land is given to the group in common ownership. The members are free to decide whether they want to divide it up and work it individually, or to run it collectively. No admission fee is charged, but each member of the group applying for land must con-

[8] Stalin, *ibid.*

tribute his share to the expenses incurred in the process of land assignment.

Quite different was the original formation of the *Kvutza*. There were no administrative measures, no stipulated requirements, and no due processes of land assignment. There was simply a small group of people devoted to the task of building a Jewish home in Palestine. After freeing themselves from the uncongenial supervision of a professional agronomist, they gradually, experimentally testing their way ahead, developed out of their own free decisions what is called the *Kvutza* or *Kibbutz*. Once a small group of pioneers had set the pattern, and others in relatively larger numbers had begun to emulate it, the formation of a *Kvutza* became formalized. Today there are two ways for joining such a settlement: one may apply for admission to an already existing group; or one may join a group which prepares for settlement. To be eligible, in both cases, one must be a Zionist, over eighteen years of age, in good health, and of good character. In the first case, one serves as a candidate for a period of six months to a year, during which time one enjoys virtually all rights of membership, except the vote. At the end of this period the case of the candidate is brought before the General Assembly, which decides about his or her admission. No admission or any other fee is paid, but the new member is expected to put all his possessions into the common purse. In the second case, the applicant takes part in a training which often begins prior to emigration to Israel, in one of the Pioneer Training Farms. The training is designed to develop the aspirant's capacity to work and live together with others aiming at the same goal. Groups thus prepared form a *nucleus (garin)* which stays together after immigration to Israel. They continue, for a longer or shorter period, their preparation, while handling all affairs communally, until such time as they are assigned land for settlement. The period from the start of preparation to final settlement sometimes lasted as long as five years. The establishment of the State of Israel made larger areas available for agricultural settlement, and the waiting period has now been shortened considerably.

The difference in the manner of origin and group formation between *Kolkhoz* and collective *Ejido* on the one hand, and the *Kvutza* on the other, has had an effect on the numerical expansion of the respective types of settlements. If the authorities succeed, in one way or another, in achieving compliance with their decrees, the establishment of such settlements can proceed apace. Where they are left to grow organically, out of the free will of the participants, their increase in numbers will be slower. In view of the dictatorial powers of the Soviet authorities it is no wonder that the *Kolkhoz* has become the prevailing form of rural organization. At the beginning of the Second World War there were some 250,000 *Kolkhozi*. All Soviet agriculture was virtually collectivized by that time. In Mexico, under Lázaro Cárdenas' sympathetic government (1934-40) the number of collective *Ejidos* rose to several thousand; it is said to be down to less than a thousand at the present. The number of *Kvutzot* and *Kibbuzim* is still not far above 200.

The *Kvutza*, however, while appearing to lag numerically far behind the other two main types of the modern cooperative farm, is far ahead of them in terms of the actual practice of cooperation.

All three types of community are, as we have seen, internally autonomous, at least in theory. But the *Kolkhoz* and the *Ejido* are much more dependent on government-controlled agencies than the *Kvutza*. The *Kolkhoz* is part of a state planned economy. Its dependence on decisions made by the state authorities, particularly the Gosplan, is obvious. More importantly, it is under direct control of the so-called Machine and Tractor Station. The latter had started as a machine-lending center, but has since become the "heart and center of the local agricultural administration." Today the M.T.S. provides the *Kolkhoz* with all large-scale machinery; its staff trains the members in the required skills, and advises them on rotation of crops, the proper use of fertilizers, soil conservation, and other related problems. Above all, the M.T.S. enforces the delivery of that part of the farm produce which the state claims as its share. All the M.T.S.'s are today run by the state. Their number rose from 158 in 1930 to some 7,000 prior to the outbreak of World War II.

Similar, though less stringent, supervision is exercised by the state over the collective *Ejido*. There are two main supervision agencies: (1) *The National Agrarian Commission* which, through state commissions, directs the establishment of the settlements; and (2) *The National Bank of Ejido Credit* which, in addition to furnishing the funds necessary for the running of the settlements, performs supervisory functions similar to those of the M.T.S. The *Ejido Bank* has been described as a combination of banker, agricultural expert, family doctor, schoolteacher, lawyer, athletic director, and personal adviser of the *Ejido*.

It is true that the *Kvutza,* too, has received both land and credit advances from the *Jewish National Fund* and the *Foundation Fund* respectively. From the moment of its formation, however, it has been essentially on its own. In its relations with administrative agencies, the role of the Kvutza has been invariably that of contract-partner rather than of controlled dependent.

More marked than any other is the difference in the extent to which cooperation determines the internal activities of the three farm types. Only large-scale agricultural production is carried on cooperatively in the *Kolkhoz* and the *Ejido*. In both, work is done by the members themselves; outside labor may be hired only in time of emergency. In the *Kolkhoz,* the members form "work-brigades" composed of five to fifty members, depending on the specific assignment given by the Executive Board. Each brigade is directed by a foreman. In the *Ejido,* work is organized less strictly, but each member must obey the orders of the elected work-chief. Indicative of the nature of the Model-Rules, governing work relations, is one forbidding members to accept any outside work so long as the *Ejido* itself is in need of their labor.

Cooperation thus limited requires a rather complicated and cumbersome method of accounting. There are two sources of income for the members of the *Kolkhoz* and the *Ejido.* One is derived from the individual sector of production still in existence: one acre or two of land, a cow, a few pigs, and so on, in the *Kolkhoz;* a few small animals, like poultry and pigs, in the *Ejido.* The main source of income, however, is large-scale, cooperatively-run agriculture. In the *Kolkhoz* and the *Ejido,* the members' share in the harvests is based on the number of labor days contributed during the year. In the *Kolkhoz* this share is calculated after deductions for taxes, reserves, construction and repairs, on the basis of a measure called "workday" *(trudoden).* The measure is supposedly both quantitative and qualitative. An unskilled laborer will require more hours than a skilled one to fill his *trudoden;* even so, the final value of the workday will depend on the total annual income of the farm, and will thus vary from year to year. In the Ejido there are three methods of compensation for work: (1) wages, differing according to skill; (2) piece rates, paid during the cotton-picking season; and (3) equal shares in the common profit. Work on community projects, school buildings, meeting halls, roads, is done without any compensation.

The more restrictive aspect of the work relations in the *Kolkhoz* and the *Ejido* is reflected in the measures needed to enforce discipline. Punishment is provided in the *Kolkhoz* for violations like failure to carry out assignments or to fulfill social obligations; for absence from work without adequate excuse; and for negligence in handling equipment and live-stock. The punishment may range from reprimand or warning to temporary suspension and fine, or even expulsion. In the *Ejido,* the severest penalty is imposed for: (1) continued lack of willingness to work under the direction of the elected authorities; (2) creating disorders; (3) agitation against the collective system; (4) robbery and other criminal offenses.

In comparison, the system of the *Kvutza* is simplicity itself. It has no use for work cards, advance wages, shares in profit; nor does it need any measures of punishment. Production, consumption, in fact, all social activities are cooperative. Everybody is trusted to work according to his best abilities, and to claim from the commonly available goods a share in accord with his needs. If a member works on the outside, his earnings go into the group's common purse. No penalty needs to be stipulated for absence from work or, for that matter, any other offense. This does not mean that violations of community rules do not occur. But they are dealt with in a spirit of "family" persuasion and admonition. Expulsions are extremely rare.

The varying degrees of cooperative practice cannot but have a different effect upon personal and social relationships in the three types of cooperative farms. The effect itself is observable in all three instances. Only in the *Kvutza,* however, has it produced clearly marked results. Just like a family, the *Kvutza* will go to any length in securing for its members all necessary care in case of sickness or invalidity; it has a standing budget for the support of its members' parents, some of whom are brought to live out their old age within the settlement. The unusually rich intellectual and recreational life of the *Kvutza* has been noted by all observers. There is hardly a week that passes without some concert or theatre performance, and virtually every *Kvutza* or *Kibbutz* owns a smaller or larger library.

Truly unique, however, is the remodelling of the most significant of social institutions, the family. In the *Kvutza,* the position of woman has in all respects become equal to that of man. Consequently, the choice of mate is here, possibly for the first time in human civilization, a matter of purely personal preference. Marriage does not affect the social status of either partner; both remain members of equal standing in the community. The family thus becomes group-centered, and its structure is largely influenced by the needs of the community. Children, their upbringing and integration into the community, are as much a concern of the group as they are an object of parental love. Virtually from the time of their birth, they are brought up by professionally-trained members, and special facilities are set aside for their education. This implies a larger amount of physical separation from their parents than is common in our society but, according to all observers, it enhances rather than diminishes the mutual affection between parents and children. The family thus cannot but even more profoundly affect the individual growing up in it. No study employing objective methods has yet been undertaken to gauge this effect with any precision. There is, however, a preponderance of opinion according to which the successful emergence of the State of Israel from the trials accompanying its birth was largely, if not altogether, owed to the heroic self-sacrifice and civic discipline engendered by the *Kvutza,* and so conspicuously displayed by its youth.

To sum up, then: we find that the *Kolkhoz,* the collective *Ejido* and the *Kvutza* have the following features in common: (1) their motive of origin, which in each case is related to the needs of agriculture, reform, reclamation, or resettlement; (2) adherence, at least in theory, to the Rochdale Principles; (3)

large-scale cooperative agricultural production; (4) internal autonomy, again in theory though not, as in the case of the *Kolkhoz* and collective *Ejido,* necessarily in practice; (5) an observable modification of the traditional social institutions. As to the features in which they are not alike, the differences between the *Kolkhoz* and the *Ejido* are only slight as compared with the more basic differences between them and the *Kvutza.* We may sum up these differences as follows: (1) while *Kolkhoz* and collective *Ejido* were established by administrative decree, the *Kvutza* came into existence by spontaneous, voluntary decision of the people concerned; (2) the settlements of the *Kolkhoz* and *Ejido* types outnumber by far the *Kvutza;* (3) *Kolkhoz* and *Ejido* are controlled by government agencies, enough to raise doubts as to their genuinely cooperative nature, while the *Kvutza* is virtually free of such control; (4) in the *Kolkhoz* and *Ejido* cooperation extends chiefly to large-scale agricultural production and only partly to consumption, while in the *Kvutza* both production and consumption are cooperative; (5) modification of social institutions may be observed in the *Kolkhoz* and *Ejido,* but only in the *Kvutza* has such modification reached a stage where we may speak, as in the case of the family, of remodelling.

Theory and Practice

To put into focus the mere striking similarities and differences of the three most important types of cooperative farms existing today, we have compared them in terms of their ideal rather than their real patterns. We have dealt with *Kolkhoz,* collective *Ejido,* and *Kvutza* as ideal types, thus omitting discussion of the discrepancies between theoretical intention and practical realization. A few words regarding these disparities are therefore in order.

If we accept the findings of Naum Jasny's voluminous study of Soviet agriculture, the contrast between theory and practice is most flagrant in the case of the *Kolkhoz*. Here, government control has affected the development of cooperative farming in such a way as to pervert the principle of cooperation into its opposite. Instead of voluntary participation, there is coercion; instead of democratic decisions by the General Assembly, there is a dictatorship by officials who are themselves only small cogs in a big administrative machine and whose election is "nothing but a farce"; instead of an increased sense of responsibility on the part of the cooperators, there is a tendency to shirk duties, to defraud the group for the sake of personal gain, even to the point of outright theft; and instead of a growth of freedom and a spirit of partnership, there is a state of affairs increasingly justifying the "analogy to serfdom." In short, an organization which started out as a voluntary and free cooperative, has become one "which exists not for the benefit of its members but primarily as a means of extracting farm products without regard to the needs of the producers themselves" and thus "lacks all normal features of a cooperative." Jasny concludes, "The mis-named *Kolkhoz* is the nutshell of a cooperative without the nut." [9]

Because of the relative inaccessibility of Soviet Russia to independent investigators, observations on the *Kolkhoz* may not be capable of immediate verification. That serious "shortcomings" of the kind Jasny enumerates actually exist can, however, hardly be doubted in the face of the internal and external evidence available. Their existence is confirmed by official pronouncements and by expert observers who can hardly be suspected of lack of objectivity.[10] It is worth noting that all the shortcomings mentioned are directly connected with the exercise of government control. Such control is naturally

[9] Naum Jasny, *The Socialized Agriculture of the U.S.S.R.* (Stanford, Stanford University Press, 1949), p. 298.

[10] See particularly the section dealing with social readjustment of N. Barou's article "Soviet Collective Farming," in: *Year Book of Agricultural Cooperation,* 1948.

strongest where the government is dictatorial, as in the Soviet Union and the countries of eastern Europe under Soviet influence. It will be found operative in the case of the collective *Ejido* as well. There, however, the distortions of the cooperative principle are less extreme than in the case of the *Kolkhoz,* apparently because of the more democratic control exerted by the Mexican Government.

Of the three instances of cooperative farming, only the *Kvutza* appears to be free of the evil consequences of government control. This does not mean that the *Kvutza* is not beset with critical issues. But they are of quite different nature. The problems facing the cooperative farms of the *Kvutza* type in Israel today arise partly from internal and partly from external changes. Internal problems, such as an increasing demand for personal comfort, lack of participation in the General Assembly, and a certain sense of frustration, particularly on the part of the women, are related partly to social and economic solidification and partly to the growth in size of the settlements. The establishment of the State of Israel and the requirements of unrestricted immigration, the absorption of the most active *Kvutza* members in state functions, a tendency on the part of the state to interfere in the internal affairs of the settlements, and a disinclination on the part of the new immigrants to join its ranks,[11] have had their effect upon the *Kvutza.* Yet serious as some of the problems are, they have not affected the genuinely cooperative nature of the *Kvutza.*

A Field of Social Research

In accordance with the prevalent mode of human relations, sociology has, thus far, devoted its chief attention to the processes of conflict and competition. Little, if any, space in current sociology text books is devoted to cooperation as more than a general "social process." And this despite the fact that

[11] "Pains of Growth in Israel" in this volume.

the cooperative movement has grown from a handful in a small English town in 1844 to a world-wide membership of some 150 millions.

This neglect may be related to the traditional academic disregard for the "small man" and his economic plight. A better explanation, however, may be found in the nature of the cooperative movement itself. So long as that movement did not extend beyond consumers' purchasing and marketing associations, cooperative practice necessarily remained segmental. It is true that by joining a consumers' cooperative an individual enters into direct cooperation with other members of the association. But he cooperates only in terms of one of his many social roles, namely that of consumer. In his other relations he not only fails to cooperate but is free to compete, even with the members of his own association. More often than not he is compelled to do so in the role of seller of goods or services. Such cooperation is too limited in itself to produce marked changes in personal attitudes and social behavior. The main motive for participation in segmental cooperatives is economic convenience. The fact of motivational restriction may, incidentally, explain the often deplored but apparently incorrigible lack of membership participation in the general affairs and conduct of cooperative societies.

Such lack of participation is hardly possible in cooperative communities, which, as we have seen, are rural settlements with cooperation as the basis of economic and social organization. In such a community the individual always cooperates as a producer, and, in an appreciable measure, in most of his other social roles as well, as a consumer, as a participant in self-government, as a seller and a purchaser of goods, and so on. As potentially, if not actually all-inclusive, cooperation of that kind affects more decisively the basic social institutions, hence all human relations within the group. Here cooperation is no longer segmental; it has produced a new, the cooperative way of life.

The cooperative community deserves the most serious consideration of the social sciences. As even our short and necessarily sketchy survey has shown, it has left behind it the "alchemistic" search for a cure-all for all the evils besetting human society and has instituted controlled experimentation in order to discover solutions for given problems of rural economy. In Palestine, in the Soviet Union, in Mexico, the cooperative community has meant, as we have seen, the introduction of the most advanced farming methods into formerly backward rural areas. Wherever the cooperative community has succeeded in establishing itself, it has brought medical care, improvement in diet, and more decent and sanitary housing to people who formerly lived in dirt and squalor and suffered from malnutrition and other diseases of poverty. As full-fledged members of a cooperative group, destitute farm folk acquired, often for the first time in their lives, a sense of economic and social security. By fostering participation in the arts, literature, and scientific progress, this type of cooperative organization has gone a long way to dispel the "idiocy of village life." There is little utopian about a type of social organization which can point to such realistic accomplishments. As a relatively small, but at the same time comprehensive and clearly defined social system, the cooperative community is a most attractive field of legitimate social research. It offers an ideal opportunity for the scientific observation of data required for the foundation of a sociology of cooperation capable of making a constructive contribution to our sorely tried civilization.

(3)

SOME RECENT DEVELOPMENTS IN
COOPERATIVE FARMING

The major developments of cooperative farming today are the *Kvutza*, and related types of cooperative communities, in Israel, the collective *Ejidos* in Mexico, and the Soviet *Kolkhoz*. Less conspicuous, but well observed and studied, are similar attempts in Saskatchewan, Canada, and the *Communities of Work* in France. They are all treated in more detail elsewhere in this volume. There are, however, a number of instances of planned or actual development in cooperative farming in other countries which thus far have received little notice. The purpose of the following notes, based on the most recent information available, is merely to call first attention to them, not to provide strictly up-to-date and statistically precise records of plans and accomplishments. Such precision, while desirable, is difficult, if not impossible, to obtain because of the lack of reliable source material, especially for countries in Eastern Europe and the Far East.[1]

Situated in the direct sphere of influence of the Soviet Union, east European countries have emulated the Soviet system of collectivized agriculture. The governments of Poland, Hungary, Bulgaria, and Czechoslovakia—Yugoslavia is a special case—have resorted to collectivization, usually against the will of their peasant populations. Information about the results is so scanty and unreliable that it would serve no useful purpose to present it here. In western Europe, cooperative farming has received only tentative attention. Its practice

[1] For information on the most recent developments in this field, see the column "Current Developments" of *Cooperative Living,* a periodical published by the Group Farming Research Institute, Poughkeepsie, N. Y.

is spotty and limited to isolated instances, as may be seen from the following observations in England, Sweden, and Cyprus.

Britain

Cooperative farming is not without historical precedent in Britain. The *Weekly News Service* of the Ministry of Agriculture (No. 149) had this to say about the pooling of resources in the countryside: "Centuries ago, when this country had to grow its own food and the population was little more than that of London today, the frequent absence of able-bodied men caused by some major or minor war, made it very difficult for the land to be kept in continuous cultivation. About the time of the Norman Conquest it was found that good farming could be maintained only by a committee in each village regulating the use of labor on the land and pooling oxen and such implements as were then available, and supporting a common herdsman."

On a small scale, cooperative farming was also resorted to, and for similar reasons, in more recent times. Before the outbreak of World War II, the Welsh Land Settlement Society, together with the Monmouthshire County Council, established several "cooperative profit-sharing farms," in order to alleviate industrial unemployment. During World War II, British conscientious objectors, like their American confrères, in search of a "moral equivalent of war," embarked on various cooperative community ventures—alas, with little practical success.

A more steady, though by no means smooth, course was charted by a cooperative community of a group of refugees from Nazi Germany. The group originated in Germany in 1920. Dr. Eberhard Arnold, a writer and leader of the *Neuwerk* branch of the Youth Movement, and his wife, Emy, were the founders of the group. It was motivated by ideas similar to those of the Hutterites, with whom it became formally affiliated in 1930. With the ascent of Hitler and the ensuing persecution of racial and religious minorities, the group moved from the village of Sannerz, not far from Frankfurt on the Main, first to Liechtenstein, and later to Britain, where in 1936 it acquired property at Ashton Keynes, Swindon, Wiltshire. After great initial difficulties, the group, which called itself the *Cotswold Bruderhof,* prospered and grew to a community of more than three hundred men, women and children of various nationalities, mainly German, Swiss, British and Dutch.

When, at the outbreak of World War II, the German refugees in Britain became subject to various official restrictions, the group decided to move again. This time, it went far overseas, to Paraguay. Only three members stayed behind to wind up the business of the community. The others reached Primavera, in East Paraguay, after a long journey and many tribulations. There they began to build anew their communal life (see further under *Paraguay*). The liquidation of affairs in Britain of the *Cotswold Bruderhof,* however, took longer than had been expected. In the meantime the three remaining members found themselves approached by several new and eager candidates. They decided to stay on and form a new group. The membership soon grew to more than twenty, moved to Lower Bromdon Farm, Burwarton Bridgnorth, in Shropshire, and established the *Wheathill Bruderhof.* According to latest reports, the membership has grown to 160, including 60 children. The group practices mixed farming on 532 acres, and owns a dairy herd of 30 cows, 55 young cattle, 214 ewes, 170 lambs, and a few pigs. Other activities include: poultry-keeping, market gardening, fruit growing, carpentry, engineering, welding and cobbling. The products of the farm are used mainly in the household of the group; surplus is sold on the market.

Unlike the Hutterites in the United States and Canada, the

new *Bruderhoefe* in Britain and Paraguay avoid isolation and hold themselves open and easily accessible to the outside world. One of the first enterprises of the Paraguay community was the establishment of a clinic which serves the whole surrounding area. The British group is publishing a magazine, *The Plough,* as well as books and pamphlets to propagate its ideas, and to demonstrate their applicability to present-day problems.

As one might have expected, the coming of the Labor Government stimulated interest in agricultural cooperation. The Agricultural Act of 1947 encourages cooperative financing of smallholdings and even provides for "actual cooperative farming"; to wit: "A smallholding authority may, with the approval of the Minister, let a smallholding, or two or more smallholdings together, to persons proposing to farm the smallholding or smallholdings on a cooperative system, notwithstanding that all of the said persons have not had such experience as aforesaid, but before approving any letting under this subsection the Minister shall satisfy himself that the aggregate agriculture experience of the said persons is such as to render it likely that in cooperation they are or will become qualified to farm on their own account." [2] While this provision undoubtedly reflects progress in official thinking about cooperative farming, it can hardly be said to display enthusiasm.

Sweden

In 1945, the Swedish Government appointed a committee to draft legislation relative to cooperative farming. The impetus for such legislation had come from the favorable experience with the agricultural purchasing and marketing cooperatives. That experience suggested a further and more compre-

[2] Quoted by Margaret Digby, in: Noah Barou (ed.), *The Cooperative Movement in Labour Britian* (London, Victor Gollancz, 1948, p. 60.

hensive application of the same principle. Cooperative farming appeared feasible because most Swedish farms are too small for the efficient use of large machinery. To be sure, the co-operative principle had to be adapted to the particular tradition of individual freedom and independence of the Swedish peas-ants. To reconcile economic necessity with social exigencies, the committee, in the report it issued toward the end of 1946, recommended the organization of "joint farming societies" *(sambruksfoereninger)*. A "joint farming society" may be formed either by farmers of the same neighborhood who, as in Saskatchewan, Canada, would pool land, equipment, and other resources, or by farm laborers and others with farm experience who may rent or purchase land from the State to be worked in common. One such society was formed at Lin-koping towards the end of 1949.[3]

In view of the difficulties which the Saskatchewan Govern-ment has had with the application of the Veterans Land Act to cooperative farming, it is interesting to note that the Swed-ish law makes the "joint farming societies" eligible for loans and grants which are generally available to individual farmers. It thus sets the stage for the development of cooperative farm-ing in Sweden, although reports of the actual establishment of further "joint farming societies" have not been received thus far.

Cyprus

While serving with the British Army in the Middle East, a number of Cypriot soldiers became acquainted with the Jewish communal settlements of Israel, the *Kvutzot*. Upon their re-turn home at the end of the war, thirty of the ex-servicemen decided to establish a cooperative farm of their own. They pooled their gratuities and resettlement grants and acquired,

[3] Cf. J. W. Ames, "Cooperative Farming in Sweden," *Cooperative Living,* I, 3 (Winter 1949-50).

for the sum of £2,980, a partly abandoned farm of some 400 acres, located at Onisha, in the Kyrenia District. With the assistance of the Registrar of Cooperative Societies, they raised a total share capital of £4,450, drew up by-laws, and organized themselves into the *Onisha Cooperative Farming Society,* which was registered in July 1946.[4]

The conditions under which the men took up their new life were extremely hard. The land was poor. It had to be cleared of stones, and cleaned of weeds. Community life required readjustment and adaptation. Seven of the men could not stand the strain and withdrew. Those remaining seem to be making a success of it. They own a tractor, reaper and binder, and a threshing machine and practice mixed farming. They grow cereals and vegetables, and raise sheep and lambs, rabbits and poultry. To burn lime, they built an oil kiln which proved profitable enough to pay a third of the annual wages as well as other expenses.

The Onisha group does not practice the same degree of comprehensive cooperation as the *Kvutza.* Members deposit an admission fee and investment capital; they pay themselves wages, and share equitably any accruing profit. But as the need arises, they adopt certain features of comprehensive cooperation. For instance, to reduce their food expenses, they decided to make their meals communal.

These ex-servicemen are thus repeating in Cyprus what their Canadian confrères have attempted in the veterans cooperative farms in Saskatchewan, and apparently with like success. As the Registrar says: "As a society, Onisha is doing well. The committee works hard, the members are loyal and well disciplined, and the cooperative spirit is evident." Their example, at any rate, inspired another group of fifteen farmers, mostly ex-servicemen, to establish, in 1947, a second cooperative farm. This farm, on 80 acres of irrigated and highly

[4] *Co-operation in Cyprus,* No. 10 (September 1948).

developed land, is located at Pervolia, in the Larnaca District.

As in Saskatchewan, these cooperative farms seem to be growing in a social climate generally favorable to cooperation. There has been a steady growth of cooperative societies since. Of particular interest is the establishment of four machine sharing cooperatives. The first was founded in 1947, by 15 farmers at Yerolakkos, the second by seventeen men at the Turkish village of Kivisil, the third by 21 men in the village of Tymbou, and the fourth only recently, at Kophinou, in the Larnaca District. In addition, the Cooperative Credit Society of Lefkoniko purchased agricultural machinery valued at £2,088 and rents it out to those of its 571 members who wish to make use of it.

Paraguay

Like several other South American countries, Paraguay welcomes immigrants who are skilled farmers. Mennonites, from Russia or Canada, availed themselves of this opportunity and established a number of villages in recent years. The first cooperative community, however, came into existence when the twice displaced followers of Eberhard Arnold chose Paraguay as their final haven of refuge. (See above under Britain.)

The story of the new settlement is told in two pamphlets; *The New Bruderhof in Paraguay* (1941) and *Work and Life at the Bruderhoefe in Paraguay* (1943), published by the Society of Brothers in Britain. The story reads like a modern version of the old pioneer saga. When the advance party landed in Paraguay, it first put up its tents in the Chaco. A few months later it moved on to East Paraguay. Why? Apart from adverse climatic conditions, sparesness of population and poor transportation facilities were the principal reasons. Under these circumstances it would have been difficult for the Brothers "to go out to men" and practically impossible for men to find their way to the new community. The new settlers saw their task "not as the building of a nice Utopia just for ourselves, but as the living and proclaiming of a brotherly way of life among men."

Fortunately, a better place in a more populated and accessible region was found in East Paraguay. In Primavera, located about forty miles east of the Rio Paraguay, near the village of Itacurubi del Rosario, an estate of 20,000 acres was bought for £4,500 and called *Bruderhof Isla Margarita*. With new arrivals from Britain the group had grown to some 350 people, when it began to organize its activities. The initial difficulties were forbidding. Everything had to be started from scratch, and under the most primitive and unsanitary conditions. The land had to be broken, homes built, children cared for, and the community established with hardly any outside help.

The hardships were apt to break the spirit of less determined men. "Many were suffering from ugly tropical sores, and hobbled around in great pain. Nearly all the first group had been down with malaria, and some of them were still attacked by it from time to time. Saddest of all was the condition of the children. With only one or two exceptions, they were suffering from severe eye trouble, and nearly all of them had to be isolated from their families . . ." Despite the determined efforts of three doctors among the group's members, seven of the youngest children died during the first year.

One of the first solid houses to be built was the school. Immediately thereafter came two wooden buildings, the clinic, to serve not only the members themselves but outsiders as well, the neighboring Mennonites, and the native Paraguayans.

By the end of the first year, the group had established itself on the new land. Its two main crops were mandioca—the potato of Paraguay—and maize. It also grew rice, peanuts, sugar cane and yams as well as beans, watermelons, and kaffir (millet). In addition, it started an intensive garden and nursery for peas, carrots, cabbage, lettuce and tomatoes; it planted several hundreds of grapefruit, orange, lemon, quince, mulberry and shade trees and vines. It acquired a herd of 1,800 cattle, mostly of mixed origin and not of very good breed. Some fifty cows were selected from the main herd for milking. The group built a brickworks, set up a saw-mill, and a machine and woodworking shop. The kindergarten and the school were running smoothly. The clinic—the first and only one to service an area of 500 square miles with a population of 40,000—was staffed by three doctors and six nurses and functioned well. By Christmas, all the sick in the group had recovered from malaria and the eye disease.

Having achieved a minimum of security, the indomitable group laid the plans for further expansion. It founded, at a distance of four miles from Isla Margarita, a second *Bruder-hof, Loma Hobby.*

According to latest reports received the two communities now house over 700 members, children including.[5] Conditions are still quite primitive. But the communities keep expanding. New members are being accepted, among them native Para-guayans. New branches of farming are being added. The scope of outside activities is being steadily enlarged. Contact with the group in Britain is kept up through an interchange of members.

In sum, in their own remarkable way these high-spirited pioneers are carrying out a demonstration of great significance. Maintaining all the basic features of the Hutterite colonies, they have added a new chapter to the history of resettlement. Their achievements prove that the cooperative community is the type of settlement best suited to the pioneer task of break-ing the ground in a new and undeveloped country. Motivated

[5] See the series by Will Marchant, *Cooperative Living,* III, 2 and 3, and IV, 1.

by a great and inspiring ideal, they have demonstrated that the resettlement cooperative community can make international cooperation a matter of direct action and thus become "an efficient embassy of good will among men."

India

Despite various attempts during the last fifty years to raise the productivity level of Indian agriculture, it has remained, on the whole, very low. The need to improve the general dietary standards and to supply the growing industries with agricultural raw materials has made an increase of production very urgent. One of the main obstacles is the smallness of individual holdings. The responsible authorities are, therefore, focusing their attention on the possibilities of cooperative farming. To explore such possibilities, a number of experiments have been undertaken since the end of World War II. A survey designed to assay an evaluation of these first experiments was made by the Agricultural Credit Department of the Reserve Bank of India. Its results were published, in December 1949, in a monograph entitled *Cooperative Farming,* with a Foreword by the bank's Deputy Governor, M. G. Mehkri.

The monograph opens with a concise but comprehensive survey of cooperative farming in other countries, based largely on this author's *Cooperative Communities at Work.* There follows a series of reports on the progress of the experiments as a basis for a concluding discussion of the factors that will decide the success or failure of cooperative farming in India. Of special interest to us is the part containing the progress reports themselves. It points out that, although cooperative farming was not altogether unknown before, it received wider attention when, at the end of World War II, discharged soldiers demanded better opportunity for land settlement. Partition and the influx of refugees in need of resettlement constituted an added impetus. A general factor in the initiation of the experiments was, further, the depletion of foreign exchange resources, and the consequent restriction of food imports. It became necessary to counterbalance this restriction by increased home production. And again, one way to enlarge the scale of farm operations was the pooling of resources.

Most of the experiments are conducted on land previously abandoned or waste, and brought under cultivation by new settlers. Pooling of land and resources by farm owners is relatively rare. As revealed by the following brief recapitulation of the data collected in the survey province by province, virtually all of the projects are as yet in a stage that makes it impossible, all official optimism notwithstanding, to estimate their potential soundness. Only instances of actual cooperative farming are considered here. The data are complemented with additional information culled from *Cooperative News Digest,* issued in mimeographed form, periodically, beginning with April 1950, by the Reserve Bank of India.

1—*Bombay.* According to a scheme adopted by the Government in 1947, the following three types of cooperative societies were to be sponsored: (a) cooperative joint farming; (b) cooperative tenant farming; (c) cooperative collective farming. The plans foresaw the establishment of altogether 112 farms of one type or another by 1952. At the time of the survey only 37 farms actually were at work. The reports give no indication of the type of organization adopted. It is stated merely that the farms were subsidized in line with provisions laid down in several administrative acts. Of the 37 farms, 16 had only just begun operations. The 21 older societies had a membership of 668, a working capital of Rs.2,35 lakhs (or 235,000 rupees, a rupee being equivalent to about 20 cents), and owned a total of 4,951 acres of land, of which 3,651 were under cultivation.

According to the June 1950 issue of the *Digest,* three of the farms located in this province have gone furthest in the pooling of their resources. These farms are: *Shadifalia, Ambach* and *Kohanvad*. They not only work their land in common but have also pooled their agricultural implements and livestock. They distribute income according to labor contributed. These farms appear to have succeeded well enough to have been singled out for praise by the Finance Minister of the Province.

2—*Bihar*. Joint cultivation, practiced in eight cooperative farms, concentrates exclusively on sugar cane. The oldest is the *Bithauli Cooperative Farm*, established in May 1945, with a membership of 44 that subsequently increased to 55. The society cultivates a total area of 45.76 acres, divided into 145 plots, with an average size of 0.32 acres. The farm is not subsidized by the Government but by private enterprise, the Sitalpur factory. The loan of Rs.5,000, used for purchase of bullocks, was repaid in three installments. Among the main difficulties facing the farms are listed: individualistic habits among the members; objection to manual labor considered below caste; lack of a sense of collective responsibility; shortage of livestock; and lack of irrigation and storing facilities. This list may be considered as characteristic of the kind of obstacles facing development of cooperative farming in India in general.

Of the other eight farms, three seem to have done very poorly so far, and the rest are said to have met with "some success."

3—*Central Provinces and Berar*. Three cooperative farming societies were organized at Akola, Imlikheda and Vihad.

The *Akola* society, founded in 1948, is an instance of a society formed by farm-owners. The eleven members who voluntarily pooled their land, altogether 116 acres, did so, to start with, for a trial period of five years. They entrusted their society with the ownership of the fields and ceded to it all rights of mortgage and sale. Although the land is not in one contiguous block, all cultivation and harvesting is done in common. The main crops are cotton, wheat and vegetables. The management of the society lies in the hands of a government-paid expert who works together with a sub-committee formed from among the members. Other supervisory personnel is offered, free of charge, by the Cooperative Agricultural Association, *Amraoti,* with which *Akola* is affiliated, and by the Agricultural Department of the province. The society pools the produce from all the land and, after deducting 25% for a depreciation and crop-stabilization fund, distributes the balance among the members in proportion to the value of the land contributed to the pool, independently of the nature of the crop.

The two other societies, *Imlikheda* and *Vihad,* were organized by refugees. *Imlikheda* society was founded in 1948. It owns a dairy plant, purchased at the cost of Rs.62,000, which handles 54 cows. While waiting for the allotment of land promised by the Government, the society runs a transport business, employing two trucks. Other activities include the sale of cloth produced by the members, and the management of a so-called ration store.

At *Vihad,* the Government has offered 3,170 acres of forest land to one hundred families of expert farmers displaced from Multon, which now belongs to Pakistan. The group, organized as a cooperative society, had cleared, by March, 1950, 200 acres of land for sowing. Each of the displaced families, the Digest reports, received a loan of Rs.5,000 (about $1,000.), a third of which represents a subsistance grant designed to tide it over the first difficulties. The rest is to be repaid to the government in twenty-two annual installments, the payments to begin when the whole area allotted to the group has been brought under cultivation. The cooperative farm owns in common a tractor, thirty-two pairs of bullocks and a number of agricultural implements.

4—*Orissa.* In this province, ex-servicemen and former civil pioneers have formed three land colonization societies which operate on land assigned to them by the Government. After reclaiming the land, the colonists are free either to divide it up among themselves, or to cultivate it jointly. The purpose of these societies is to supply the members with agricultural implements, machinery, livestock, and so on; to acquaint them with modern farming methods; and to assist them in developing village industries. Provisions also are made for cooperative sale of the products. The three societies are: (1) *The Gurang Champatimunda Society* at Angul, with 37 families, on 200 acres of reclaimed land; (2) the *Krishma Chandrapur Society* at Parlakemed, including 50 families working on 250 acres of reclaimed land; and (3) the *Ratangia Society* in Udayagiri, with 50 colonists still occupied with reclaiming the land assigned them.

5—*Assam.* The Government inaugurated, in 1948, a large-scale development of cooperative farming. The project, which goes under the name of *Misamari Cooperative Colony,* is still in its initial stage. It provides for the settlement of 250 families, in five administratively inter-connected villages. The area assigned to the project extends over nine square miles. It was taken over from the former settlers who are being paid off in annual rents and are offered an opportunity to join the colony. Most of them are said to avail themselves of the offer. Each family is to receive a block of land slightly larger than that owned by it prior to the initiation of the project. The plans foresee common cultivation of a sizable part of the land in each village, and the cooperative marketing of the produce. The income is to be distributed among the members in proportion to labor contributed.

At the time of the report, the prospective colonists were temporarily settled on a tract of land occupied during the war by American forces. The barracks and other buildings, altogether about 300, were purchased by the Government and put at the disposal of the new occupants. The colonists run a dispensary, a school, a postoffice, a consumers' store and a trading cooperative. Farm operations are prepared under the general direction of a superintendent who, assisted by two farm experts, is charged with the planning and coordination of all activities. A special plot is set aside for experiments to establish the most efficient farming methods for the area. The members receive instruction in the use of modern machinery, and are building their own houses.

6—*Delhi.* Two multi-purpose cooperatives were established by refugees to cultivate their land jointly. They are the *Suchkhand Multipurpose Cooperative Society,* and a like society at Jaithpur.

Actually, only the *Suchkhand* society, formed by 35 refugee families from Bahawalpur and Sind, in what is now Pakistan, was able to proceed with operations. With the help of two tractors, borrowed from the Ministry of Agriculture, this society is reported to have succeeded in ploughing half of the allotted 400 acres of land. The other half of the land is being cultivated by arrangement with neighbors who not only plough the land but also supply the seed for sowing. *Suchkhand* receives assistance and guidance from the Indian Cooperative Union. It was also granted, by the Prime Minister's Fund, a subsidy of Rs.10,000 to establish a dairy. The dairy is said to have produced during the first year about 175 lbs. of milk a day, for which it found a ready market in New Delhi. The management of the *Suchkhand* society lies in the hands of an executive committee, elected by the general meeting of members. The net profit—or what remains after deduction for loan repayments, the reserve fund, community services, and so on—is distributed among the members according to labor contributed.

The society at Jaithpur counts 30 member families of

refugees from Sind. It was allotted 300 acres of land and received a loan of Rs.20,000 from the Government. Operations were, however, delayed because of certain disputes among the settlers. The difficulties are said to have been overcome through the good offices of the Indian Cooperative Union, and the society was scheduled to start cultivation during 1950.

A cooperative farm, named *Nilokheri,* was established near New Delhi by refugees from the West Punjab. Thirty-five families work an area of 400 acres of land, and appear to be quite successful. As the June 150 *Digest* reports, they are running a cooperative dairy and use several tractors in their large-scale operations. Of great significance for the future of the settlement is the sinking of two tube wells, reaching a depth of 300 feet and capable of yielding more than 10,000 gallons of water per hour. The settlement is described as one of neat little cottages, well kept roads, parks, play grounds and workshops; it has a school, a community hall, a power generating station, a canteen and even a theatre of its own. It appears that this settlement is developing into a model community that might set an example for other projects. Encouraged by this success, the Cooperative Union of Delhi, under whose auspices this settlement was started, is said to consider larger settlements along similar lines.

7—*Madras.* A cooperative farming society named *Nelvoy* was established. The *Digest,* quoting the *Madras Journal of Cooperation,* calls it "a novel experiment in cooperative farming." The society, we learn, distributed part of the 400 acres of land, leased to it for a period of twenty years, among its fifty members. They are mainly agricultural laborers and *harijans,* or "sons-of-god," as Gandhi preferred to call the untouchables. The society has the support of the Government and enjoys all the benefits granted in India to land colonization societies. Each member receives a money grant which he applies to his share in the capital investment. The society is further subsidized by a free supply of seed and manure as well as by an interest-free loan for the purchase of equipment.

In another part of the province, in the Keel Kotagiri area, adjoining the Coimbatore District, close to the tip of the peninsula, the government allotted 300 acres of land fit for paddy cultivation to a group of farmers who organized themselves cooperatively. They did so well during the first year of operations that the Government was ready to assign them, as the July 1950 *Digest* states, another 100 acres of land. Progress is being made in setting up a Land Colonization Society to include all those interested in cooperative farming in this area.

8—*United Provinces*. According to the May 1950 issue of the *Digest,* cooperative farming is tried in two villages, one in the Jhansi district and the other in the Ganga Khadir colony of the Meerut district. Supported by government grants and loans, the farmers of the villages chosen for the experiment have pooled their land and cultivate it in common. The results are said to be so encouraging as to warrant the extension of the experiments to other districts in the Uttar Pradesh and in the Bundelkhand area.

9—*Mysore*. The July 1950 issue of the *Digest* names the following five cooperative farming societies organized in Mysore: (1) *Archell;* (2) *Vonakemaradi;* (3) *Uddur Aspathre Kaval;* (4) *Kumbhapur;* and (5) the *Madukar Hosahal Better Farming Cooperative Society*. No details are given beyond the fact that the first Management Committee was nominated by the Registrar, and that agricultural inspectors were appointed as ex-officio secretaries for the first three societies mentioned.

Ceylon

Five *Joint Farming Societies* have been established here. Two, unnamed, in the Kalutara District; one, *Kalametiya,* at the Ambalantota lagoon, another, *Horawinna,* in West Girawa Pattu; and a fifth, *Oluvil,* in Akharaipattu, in the Batticalia District.

The *Administration Report on the Working of Cooperative Societies* (January 1946) finds that the "success of these co-operative farms has been so intermittent as to make the prospects of their survival doubtful." This, partly because of adverse material conditions, but partly also because of "the result of local enthusiasms, which have too often proved transitory, if not altogether illusory."

The most successful of the five farms appears to be *Oluvil.* Its membership consists of 58 farmers who cultivate jointly about 600 acres of land. *Oluvil* had a good harvest during the first year of its existence which was owed apparently to the fact that it "introduced its members for the first time to the use of the light plough and of artificial fertilizers." Encouraged by their success, the members plan to purchase a tractor "for joint use with other similar societies that are in prospect in the neighborhood."

The experience of the five *Joint Farming Societies* leads the Report to conclude that "Cooperative farming, if it is to be successful and gain a hold, must have its merits demonstrated to the people by the State." Plans to do so are already being laid. The Ministry of Agriculture and Lands has accepted the recommendation of a Sub-Committee that settlement on the recently acquired *Ma Oya* and *Mount Mary Estates* (near Kundy) "should proceed along lines of collective farming." The results of the experiment—as the Report rightly points out—will be worth watching.

It may be added that these experiments are part of a co-operative movement which has grown considerably in recent years. One notable sign of this growth was the establishment, in 1945, of a Department of Cooperative Development. G. De

Soyza, the Commissioner in charge of the Department, signs as the author of the above report.

Pakistan

In the August 1950 issue of the *Review of International Cooperation,* H. S. K. Lodi, of the Cooperation and Marketing Department in Karachi, observes that the cooperative movement has made considerable progress in Pakistan in the two years since partition. The number of cooperatives has grown to 50,000 and their membership to 2,250,000 families, or to 13.5% of the total population. Of interest to us is the development of cooperative farming which was introduced in Pakistan only after partition. There are two main areas in which cooperative farming societies have been established: West Punjab and East Bengal. The main purpose of the societies in West Punjab is rehabilitation of the refugees from India. The societies facilitate their resettlement by providing the newcomers with land, seeds, and agricultural equipment, including heavy machinery. Each society organizes its own religious service, primary education, first aid clinic, and secures the services of a veterinary. The land is apparently worked individually, but the harvests are pooled and distributed according to a formula which divides the produce into six parts, two for taxes and reserves respectively, and the rest for the members. The marketable surplus is sold by the society collectively. Two hundred societies have been established so far. They cultivate an area of more than 2,000 acres of mainly virgin soil.

Of a more comprehensive nature are the three societies in East Bengal. They appear to be organized along the lines of the Soviet *Kolkhoz,* or the cooperative farms *(Meshek-Shitufi)* of Israel. All large-scale mechanized farming is done collectively. In addition, each family owns, close to its own house, a plot of land on which it may grow vegetables, cultivate fruit trees, and keep a limited number of livestock. The income from cooperative farming is distributed, after deduction for expenses and community services, according to the labor contributed by each family. The income from his own plot of land is at the free disposal of each member. The results of the first experiments appear to be encouraging enough, according to Mr. Lodi, to warrant the introduction of similar cooperative farms in other Provinces of Pakistan.

New Zealand

The cooperative movement in New Zealand, like that of the U.S.A., appears to be concentrated mainly in the fields of agricultural marketing, purchasing, and, above all, processing. Also of some significance seem to be the building and loan societies and investment cooperatives. Some small-scale industrial and service cooperatives have been set up by ex-servicemen since the end of World War II, with the help of the Government Rehabilitation Board.

Only since the end of the war—according to a report in the August 1950 issue of the *Review of International Cooperation*—have attempts been made to establish cooperative farms in New Zealand. These ventures were initiated by veterans who, while serving in the Middle East, had opportunity to become acquainted with the agricultural communes in Israel. The report finds that the cooperative farms met "with almost complete lack of success," although "one or two cases, involving a few close friends or relatives," have managed to survive. To the fortunate latter belongs the *Riverside Community,* located not far from the town of Moutere. This community was formed in 1948 by a group of 27 New Zealand Christian Pacifists who acquired 450 acres of land.[6]

---

[6] Cf. A. C. Barrington, "The Riverside Community," *Cooperative Living,* II, 2 (Winter 1950-51).

(4)

THE LINK WITH THE PAST

The R.S.I. (now the Group Farming Research Institute) sent a small party in the late spring of 1946 to study the development of the recent but steady growth of cooperative farming in Saskatchewan, Canada. Instead of going straight up, we decided to follow a route that would lead us through some of the cooperative or cooperative-minded communities in the United States.

The Penn-Craft Community

We had driven most of the day through Pennsylvania's steel and coal districts, through cities covered with the ugly patina of grime, past pits and factories that looked tiny besides their ghastly mountains of slack, across "company-towns" with their blackened shacks and their benighted populace. Now, at a turn of the road, a different view opened before us: a wide-stretched green valley with a neatly laid out, fair-sized settlement in its center. No soot here, no heaps of slack, no murderous slum, but gently rolling, velvety slopes instead, clusters of trees and bushes, and the beauty of a friendly countryside.

The contrast could not have been sharper. At a glance, there came to life for us the full meaning of the coldly descriptive terms: "Experimental Community for the Rehabilitation of Unemployed and Partially Employed Coal Miners." The story of the settlement went back to the Depression which to the coal-miners was only one more drastic incident in the perennial vacillation between full employment and lay-off, characteristic of their occupation. America's answer to the Depression, the New Deal, had given currency to the term "rehabilitation." This idea fitted well into the experience of

66

the American Friends' Service Committee with its long practice of helping people to help themselves. A study made in
the early Thirties showed that technological and other changes
in the coal mining industry would leave some 200,000 miners
permanently un- or under-employed. Relief was all these people could hope for. But the Friends saw another possibility,
and they decided to "experiment." They thought that the
miners could be made self-supporting if (1) they were given a
chance to produce as much as possible of their own food, and
(2) if they were retrained in cash-yielding skills that could be
practiced at home. Both these conditions, the Friends felt,
could best be achieved by group action. The establishment
of a settlement was decided upon that would "encourage initiative and the assumption of responsibility by fostering a wholesome type of community life, the encouragement of cooperative projects and the promotion of the democratic process in
community relationships." [1] An agency, Friends' Service, Inc.,
was established; money was raised; and a suitable place was
found in Fayette County, near Uniontown, Pennsylvania. Next,
an experienced community planner was engaged to lay out
the settlement; a man was put in charge of selecting the
applicants; and by 1937 the first of the fifty families, chosen
from two hundred applicants, began to build their homes in
what was to be the Penn-Craft Community.

All the available printed material on Penn-Craft[2] reveals
the unusual care taken, and the good sense used in planning
the community. The principle of helping the miners to help
themselves prevailed from the very beginning. It was to be
the miners' community, so they were to build the houses them-

---

[1] *Progress Report on Development of Penn-Craft Community
1937-1940* (Philadelphia, American Friends' Service Committee,
1941), p. 2.
[2] Frederick L. W. Richardson, Jr., "Community Resettlement in
a Depressed Coal Region," *Journal of Applied Anthropology,* Octo-
ber-December, 1941.

selves. Provisional dwellings were first erected—later to be used as chicken coops—to house the miners and their families until they could move into their permanent homes. Credit was offered to each family to the amount of $2,000, but each member was expected to put in labor equal in value to the sum advanced. A "man hour system of labor exchange" was established. This enabled the men to accumulate "credit hours" by working on their own or their neighbor's houses. The homes we saw in the valley below represented thus the results of truly cooperative action.

However, the war had brought an unforeseen boom to the mines. Employment had been steady ever since, the wages relatively high, and the rehabilitation enterprise had not gone beyond the first, the house-building stage. Did this mean that all other intentions of the planners had been forgotten? And if so, was Penn-Craft to be called a failure?

If open-hearted welcome offered strangers is a sign of healthy community, then Penn-Craft could by no means be thought of as having failed those who planned it. Passing some of the trim, solidly built, generously spaced stone brick houses, we stopped at the cooperative store. It was after store-hours, but we had a look at it from the outside: it seemed to be fairly large, well kept and stocked, and there was a locker plant attached to it. There were children playing in front of the store, and they directed us to the house of the community manager, Mr. David Day. He happened to be out, but his wife made us at once feel at home. She showed us around the house, six airy, comfortably furnished rooms, a modern tiled bathroom, a well-equipped kitchen, and an open porch. We talked about the community, and it was clear that she felt good about being part of it. Before we knew it, we were invited to stay over night. She herself could not place more than one of us; she had a boarder, a college student who had come to work and to observe. But she called Mrs. Peckham, and within a few minutes we were settled for the night and for dinner.

We decided to stroll through the settlement. Now, the first impressions over, we noticed some less attractive aspects in the appearance of the community: none of the grounds—there was more than an acre of land around each house—seemed to have ever been worked; the community center, the only old building in the place, was in a state that could be called anything but tidy; some of the children we met did not look as strong and healthy as they should have; and the roads were not paved and were quite neglected.

After a time, Mr. Day caught up with us and, while driving with him to the barn where he kept his cow, we got from him "the story." The consumers' store was all there really was of cooperative enterprise, though there were community activities, such as the Mothers' Club, with its monthly meetings, a 4H Club, a Boy Scout group, and on the whole it could be said there was more feeling of belonging than in other American communities. When two men died before their houses were finished, the others chipped in and helped readying the homes for the bereaved families. There was no discrimination as to creed or race at Penn-Craft; five Negro families lived here on equal footing with the rest of the community. Originally, Penn-Craft was built to house fifty families. With the return of veterans the number had risen to eighty. The "chicken coops" were used to house the overflow. To relieve the congestion, a new section had been laid out for the young ones. The two and a half acres allotted each family had done no good: the land was not large enough for real farming, and too large for part-time handling. The younger ones wanted to take up dairy-farming or fruit growing. The sixteen new units were planned with ten acres of land to each house. Of fifty applicants, thirteen had already been accepted.

Physically, the community was growing. There had been difficulties. With the boom in the mines, hardly any advantage was taken of the facilities to produce part of subsistence or cash-income in the community. The land lay fallow, and the "undivided land," reserved for cooperative farming, was rented out to one of the members, who kept twenty-four cows on it. The knitting factory had been a real headache. Now it was run as a private enterprise by a Viennese refugee couple who had brought experience and technical skill with them. They employed seventy workers, among them only four community members.

The people got along quite well with each other. The only noticeable friction arose from the racial equality practiced. It came up in the social affairs of the youth. Often a neighborhood boy would refuse to attend when he learned that there was to be a Negro boy at the party.

Thus, all that really had succeeded at Penn-Craft was the housing. Still, the achievement was by no means small. People here lived together under more favorable conditions than neighbors around them. They had developed a sense of responsibility, and were beginning to think in terms of cooperative action. We received a sample of their concern for the community when, after dinner, a group of them came over to Mrs. Peckham's home and involved us in a serious discussion about adult education.

There was something extremely hopeful about this community. These simple miners had demonstrated to themselves that, if given a chance, they could improve their lot through their own action. They had helped themselves to houses and now they were looking forward to learn the ways of helping each other to grow into a well-knit community. Was it perhaps that cooperative living in America, if it ever was to come, could be brought about only in this slow way and step by step?

Young Planners at Yellow Springs

Yellow Springs is a small town, but it offers several points of interest. It is best known, of course, for Antioch College, with its unorthodox curriculum of academic study and practical work. It is the seat of *Community Service, Inc.,* an agency created by Arthur E. Morgan, a former president of Antioch College, even better known for his connection with the T.V.A. This agency's purpose is to serve as a clearing center of information "on the background and principles of community life and organization"; it is quite active and publishes books and pamphlets on the subject; it holds annual conferences, and publishes a bulletin, *Community Service News.* There is also the Antioch Bookplate Company, started in 1926 by A. E. Morgan's son Ernst, then still at College, together with a fellow student as part of the program of "alternating work and study." It has grown since into a very successful business enterprise, of interest to the student of labor-management cooperation because of its interracial hiring and profit-sharing policies.

Our own reason for stopping here was, however, a group of about ten former and present students of Antioch who were planning to establish a cooperative community. The way they had gone about it, so far, indicated more than usual determination. The summer before, they had gone to North Dakota to get some preliminary training. They had lived together in an abandoned freight car and had earned their living by working for the farmers in that area. When I last visited them, March, 1946, I found them still in what might be called the "groping stage." They had not clarified their ideas to the point of definite action. The eight or ten who had shared the North Dakota experience formed a not too well-knit nucleus of a larger, loosely defined group of about twenty-five boys and girls who were interested in the idea of cooperative community without declaring themselves actively for it. They planned to

settle on land, and to build their community around a Folk-
School. They hoped to derive cash-income from handicraft,
and Dave Orcutt, in some ways their leader, was hard at work
on a novel kind of puppets whose production he hoped to
make the initial commercial enterprise of the group.

Our two-day stay was somewhat marred by the fact that
three of the most active members were absent: Warren and
Judy were going to be married in their hometown, and Dave
was on his way there, hitch-hiking, to attend at the wedding
as best man. We gained the impression that some progress had
been made. Jean and Harold, who until now had been the
only married couple in the group, had rented for fifteen dollars
a month an abandoned farm at a distance of three miles from
Yellow Springs and had moved out there. The house they
lived in was in an advanced state of decay, and the land,
neglected for many years, would not yield more than a patch
of vegetables. But they felt good about it: the first step had
been made, they were "out on the land" at last. They shared
the house with Roger, at thirty the oldest in the group, and
its hardest worker. He had obtained, on the other side of
Yellow Springs, the lease of three acres of good land, and he
worked it, practically with his bare hands.

The group gathered often in the evening on the "farm," but
met regularly for meals in the couple's former town apartment
which now served as group center and guest house. The meals
were the only undertaking in which the whole group cooper-
ated: each contributed his share of expenses, and took his turn
in shopping, preparing and serving. The arrangement was in a
sense that of a communal kitchen, but anybody interested in
cooperating could participate.

We had several meetings with the group. Their main concern at the time was how to demonstrate—to themselves as
much as to others—the need for a cooperative community.
The youth they came in contact with were content with things

as they were. They did not mind at all the ordinary routine of college and good job. Since jobs were plentiful, and wages high, what need was there for any change?

We had to admit that, under the circumstances, this attitude had to be expected. But, we agreed, it could be assumed only by the short-sighted. After all, atomic energy had been released and it could easily be shown that from now on cooperation was destined to play an ever increasing part in our social order —and this independent of whether the new energy was to be used for the pulverization of cities or the benefit of mankind.

Take the first case. Obviously there was no defense against atomic attack but dispersion. The fear of obliteration may well induce people to move away from the cities. But once they felt secure again, they would begin to miss the stimulation of the urban centers and risk anything rather than die of boredom. To make decentralization effective life had to be made attractive in the small groups out in the country. But there is no other tested means of combining the advantages of urbanity with the amenities of the country than the cooperative community and all that it implies.

Assume that there was going to be no more war. If a tiny pile of ore could produce a virtually unlimited amount of energy, the power needed for industrial purposes would become as common and general as water is today. Any overcrowding in one place would lose its main justification: easier energy supply and service. The small community would then become the given type of settlement. Abundant energy would produce an economy of plenty: by merely working according to his capacity each would secure naturally all he needed. Which again amounts to the way practiced by the cooperative community, particularly the Hutterites and the Israeli *kvutza*.

Our argument seemed convincing to those who were in on the discussion. We all agreed that, although possibly isolated now, those who took up the idea of cooperative community might find themselves generally accepted before long. There remained only one doubt in our own mind: could cooperative community be built on reason alone?

Before leaving Yellow Springs, we paid a visit to Arthur E. Morgan, mainly to inquire about the state of affairs at the *Celo Community*. This project had been started a few years ago by Mr. Morgan and some of his friends. It was located on a tract of 1,200 acres of land in North Carolina, about fifty miles northeast of Asheville, and was to be conducted in a strictly scientific manner. The carefully selected settlers were left free to develop any social organization they thought suitable. We learned that there were eight families living now at *Celo*. As at Penn-Craft, high wages on the outside kept the people from developing their subsistence basis, which was to be chiefly farming. All they were doing at present was preparing the soil for the time when they would want to work it. The one communal activity was the erection of a Health Center that would house a clinic of six to eight beds and offer Health Training facilities. Scarcity of building material was delaying the completion of the Health Center.

Though candid about the paucity of *Celo's* achievements, Mr. Morgan was nevertheless confident about its future. Even though the progress was slow, we agreed with him that the experiment was worth watching. But we could not help wondering whether the organizational abstinence had not its dangers. It was certainly sensible to keep away from blueprints. But was it equally wise to abandon all specification of goals? How could one estimate progress or failure if no point of reference was established? It seemed to us as if this was driving the "scientific" attitude somewhat too far.

Amana, the Community of True Inspiration

The seven villages of the society lie off the Highway. But

we could hardly miss them: for either way you come driving on U.S. Highway 6, a Gas Station & Sandwich Shop, run by Amana, is there to signal you.

The thing that is today the *Amana Cooperative Corporation* is the last step in a development which began more than two centuries ago in Germany, in the Province of Hesse. From its inception, and as long as it remained in Germany, the group of peasants and artisans was not distinguished by any but its religious tenets. They believed that true inspiration was not a matter of the past; God still could reveal himself to man, and did so through the group's leaders, his chosen *Werkzeuge,* or Instruments. Persecuted for its resistance to military service, the group decided to emigrate to the New World. It was when, 800 strong, they settled in 1843 near Buffalo, New York, that under the leadership of Christian Metz they organized themselves into a cooperative community. They maintained this type of organization after moving, in 1855, to their present location in Iowa. In 1932, facing economic difficulties, the group which had grown to over a 1,000 decided to relinquish the extreme kind of cooperation it had practiced for over ninety years, and to form instead a Joint-Stock Corporation. This "Great Change" did not fail to attract attention, and to be widely hailed as proof that cooperation could not work.

In those who have known the beauty of the German countryside, *Amana* is apt to evoke nostalgic reminiscences. It is laid out in a manner typical for the German *Strassendorf,* or long stretched village, of rectangular, solidly built houses covered with vines and surrounded by flower bushes and fruit trees. In 1946, a year after Hitler's death, that nostalgia is likely to be marred by memories of Belsen and Buchenwald. But after all, these people were here because their ancestors had rejected German militarism.

We stopped at the Ox Yoke Inn. Before the Change, this had been one of the communal dining halls; now, it had been converted into a quaintly furnished eating place for tourists. It was here that we were given the first impression of what was on these people's mind. To break the ice with our waitress—a pretty, well-built, blond girl—I addressed her in German. Although somewhat reserved at first, she seemed pleased and answered in the same language which she used without the trace of a dialect. We soon learned that her father was a mechanic and that he worked in the Refrigerator Factory. She herself was one of three children; a sister was employed in the Wool Mill, and a brother was still in the Air Force. She had gone through high school where, like in all schools of *Amana,* the language was English; but at home they spoke mostly German. There were no movies in the villages but she went to see them in the nearby town. She enjoyed going out to parties and affairs in the neighborhood. When I asked whether she did so with boys from outside, there was a revealing indignation in her statement: "Sure, we all do: we are no different from any other community."

After lunch, we walked through the village and stopped at one of the stores; its owner had been recommended to us as the best source of information on *Amana.* He was an "old-timer" who had taken a leading part in the Great Change, and was quite willing to give us his point of view. He believed that the kind of life *Amana* had led before had, by its very nature, been doomed to failure. According to him, cooperation had worked only so long as the religious beliefs retained their hold over the people. When their faith weakened—partly because no new Instruments arose, partly because of the impact of the outside world—human nature proved incapable of sustained effort without the incentive of profit. The shirkers increased in numbers, outside labor had to be hired in excessive numbers, and when the Depression hit farming, there was only one thing that could save the society: introduction of free enterprise. He was satisfied that the essential ideals of *Amana* had not been abandoned; but that the Change had made the society economically sounder than it ever had been.

We strolled on to the Central Office of the Corporation and found here the "new spirit" in undisputed reign. The gentleman to whom we spoke was obviously out to impress us. Unfortunately, we were a poor audience for the show he put on. To us what was intended as a brisk efficiency act had the rattling ring of good old German arrogance. When we asked how things were going he answered—not without a grunt of contempt directed at those "bureaucrats"—business would be all right if only the OPA and its price controls would be done away with. We wondered whether any study had ever been made of *Amana*. He threw his arm out in a gesture of disgust: he was against it, not that they had anything to hide; it would simply be too much bother. He obviously did not like to be asked questions, and to end the interview he handed us an official, illustrated pamphlet called "The Industrious Amana" which, he said, contained all the information we needed. Glancing at it, as soon as we had left him, we found on the first page such startling statements as: "The Amana Society was originally started with a profound religious and spiritual motive in a quest *for the American way of freedom of belief and worship.*" Or, even better: "Today, owning their land and factories in common as stockholders in a Two Million Dollar Corporation, spurred by a common concern for the group interest by the individual desire to excel in competition, and motivated by two centuries of brotherly love and neighborly cooperation, *these straight thinking, fine American men and women* of the Amana Society have accomplished a noteworthy achievement in economic soundness and success." (Italics in both quotations mine.) Such statements were certainly apt to arrest the desire for further inquiry.

It was only in the evening that we were able to get a glimpse at the other side of the picture. We had made the acquaintance

of one of the *Amana* families who invited us to their home,
after dinner. This family consisted of a very old lady—probably one of the oldest living members of the Society—her
daughter and son-in-law, and a granddaughter of about five.
In our conversation with the store-owner we had also touched
upon the relation of *Amana* to Germany; he had firmly stated
that all ties with the land of origin had been completely cut
long ago. One of the first things the old lady was anxious to
show us was an old Inspirationist book in German, dated
1758, and entitled: *Gericht über die Stadt Berlin* (The Day of
Judgement For The City of Berlin). We assumed she did so
because they felt like we did that the punishment this city had
received only a year ago was well deserved. But when we
referred to Hitler, we were surprised to hear the younger lady
say: "Oh Hitler, we all here were for Hitler. We felt Germany
had been treated badly after the last war—you know, the
Versailles Treaty and all that—and we believed that Hitler
would make Germany great again." When we mentioned the
concentration camps, the gas chambers, and the systematic
extermination of millions by the Nazis, they were genuinely
shocked. They were too honest themselves to doubt our words;
they had heard about these things vaguely before, but they
were made to think that these were merely the usual atrocity
stories.

When the conversation turned to the Great Change, the
younger lady asked me to read a letter which she had received
recently from the *Bruderhof* in Paraguay. When Dr. Eberhard
Arnold, the founder of the *Bruderhof,* came to *Amana* in 1931
—on his way to the Hutterites with whom his group, formed
in Germany after the First World War, later affiliated—she
had met him in the communal kitchen where she worked at
that time. She kept up correspondence with him after his
return to Germany, and later, after his death, and after the
group, expelled by Hitler moved first to England and from

there to Paraguay, with some of the members. The letter contained interesting figures on the growth of the new settlement; it owned now 20,000 acres of land, and 2,200 head of cattle; and it had grown in membership to 400. But this was not why I was shown the letter; it was rather because it contained a paragraph describing the distress felt in that far-away group about *Amana's* Great Change. They were deeply shocked about this betrayal of the common ideal.

This sentiment seemed to be shared, at least, partly, by this *Amana* family. They did not appear to have accepted as happily as the pamphlet insinuated their role as "straight thinking," profit-sharing Americans. They admitted that the discipline imposed by the Elders of the Community in former days had often been too strict; but even so, it had not been easy to accept the change. They themselves had been on the side of those who proposed migration instead. If they moved to an isolated place, say Alaska—they thought—they could escape the evil influences of America which were the cause of all troubles. But there was no Instrument to guide them, and the "others" won out. So, they were well off now, with business booming, and all that. Yet, there was that prophecy in the old books which said that if *Amana* ever deviated from its traditional ways, it would return to them after a short while. They could not help hoping that it would come true.

After duly visiting the Woolen Mills, outside of *Amana;* and the impressive Refrigerator Works, in *Middle Amana;* and after spending the night in the friendly, but primitive, hotel at Homestead—the only one of the villages not founded by the Society but purchased as a whole because of its usefulness as a railway station—we were again on our way. We left *Amana* with definitely mixed feelings. The official briskness on the one side, and the nostalgia for the old days of faithful we-feeling on the other, did not give us the impression that all was well with this Community. These people were materially better off than before; but did this mean that their life made more sense?

An American Folk School and the Bass Lake Farm

Our main and last stop before entering Canada was to be the Hutterite colonies in South Dakota; we reached them after three short, but not unwelcome delays.

The first of these was caused by our visit to Danebod, a suburb of Tyler, Minnesota, where one of the several Danish Folk Schools in the U.S.A. had existed for more than forty years. It had been closed in 1931, but the building was still there, and attempts were being made to open it again, in a newer version, as an *American Folk School*. Those who propagated the idea maintained that America was badly in need of this kind of Adult Education. It had proven effective in training the Danes in cooperative ways; it should be able to do the same with American farmers. Theoretically at least, this seemed to be plausible enough. But as we learned from the hospitable Pastor Enoc Mortensen of Danebod, there was no agreement among the proponents about the concrete purpose to be served by the American Folk School. It could not be the same here, where school attendance continued until eighteen, as in Denmark, where formal education ended at fourteen. But what else could it be? Leaving a final solution of this question in abeyance, Reverend Mortenson tried to do his best by offering in the refitted building of the old *Dane-Hojskole* summer short-courses "in cultural and cooperative living, not only for young people, but for all age groups." Later on, he hoped, a true American Folk School Movement would develop from these cautious beginnings.

Further on, at Detroit Lakes, we came across a summer camp held for a group of about forty boys and girls of high school age by the Farmers' Union. We spent a worthwhile evening talking to the young campers. We were impressed by their genuine interest in cooperative farming. Their questions showed that more than one of them had given quite a bit of thought to this problem. Of course, they were the sons and daughters of the less prosperous and more progressive-minded farmers. But after all, if this innovation ever came to the U.S., it would have to get a start in this section of the farm population. It was gratifying to see the more articulate of these future farmers take such a definite stand in favor of cooperation as a way of living.

Finally, we stopped at the Bass Lake Farm. This visit was less encouraging. The spirit of failure pervaded this small group of some ten men and women, and to all appearances the project was in the process of dissolution.

This cooperative farm had been founded, in 1941, by an Episcopalian minister, Rev. Marston, and two fellow pacifists. They bought the farm of eighty acres, located at Bass Lake, near Minneapolis, for $6,000. They added some cattle and chickens, and made a fair go of it. Then came the war. The two lay-pacifists resisted the draft, and were placed in a Civilian Public Service Camp. Rev. Marston, left alone, kept the farm going. With the war over, the two resisters returned to the farm; they brought some others with them, and still others drifted in later on. Now, they were running the farm in common, on a profit-sharing, wage-paying basis. They lived in one crowded building, but each kept house for himself.

The reasons for the breaking up were given me by Rev. Marston on a walk through the farm. He put the blame on two things mainly: the women, and lack of familiarity with farming. Some of the women were from the South and wanted to return there because they did not like the colder climate. Born and raised in Minnesota, Rev. Marston himself was not disposed to leave the farm, which would have meant also to abandon his very satisfactory ministry. More crucial still was the lack of farming skills among the members. An experienced farmer himself, he had undertaken to teach the others what he knew. But it was a tiresome job: they did not seem able or willing to get ahead with their work. So, for both reasons, it seemed best to call it quits.

The mental state of the group was revealed to me in a talk with one of the members. He saw himself as a "religious visionary" who found cooperative community conducive to his calling. He intended to devote himself to the "contemplation of the roots of humanity's evils." I wondered about the use to which he would put his findings. But he refused to consider any practical application. I tried to point out, cautiously, how valuable any revelations of this kind would be to our ailing times. But I did not really intend to embarrass him, and I was satisfied to let it go with his excuse that he was a poor teacher.

A few months after our visit we learned that Rev. Marston, who was in his early forties, had died, suddenly, from a heart-attack; the members dispersed shortly after. The last we heard was that some of them were about to join the *Celo Community*.

The Hutterische Gemein'

Our trail—long in terms of the distances covered, but only too short on cooperative community finds—was to come to an end in the two Hutterite communities of *Jamesville* and *Bonhomme*, both in South Dakota. We looked forward to this visit with a great deal of expectation. A large part of the first field study of our Institute had been devoted to the study of the Hutterites. J. W. Eaton, then research director of the Institute, and S. M. Katz, his associate, had spent a whole month at *Jamesville*. They had brought home a wealth of material, later used in several publications.[3] We had, there-

[3] Joseph W. Eaton, *Exploring Tomorrow's Agriculture* (New York, Harper and Brothers, 1943), chap. xxvi, Henrik F. Infield, *Cooperative Communities at Work* (New York, The Dryden Press, 1945),

fore, some advance knowledge about these unique survivals of religious utopian communitarianism whose beginnings went back to early sixteenth century.

It was, to be exact, in 1526 when a motley crowd of Swiss, Bavarian, and Tyrolian Anabaptists fled before the Peasants' War to Moravia because of their belief in nonviolence. Here, under the leadership of Jacob Hutter—from whom they take their name—they adopted the brotherly form of equality. By doing so they meant to emulate Christ and his disciples of whom it is said, in Acts II:44, 45: "And all that believed were together, and had all things in common; and sold their possessions and goods, and parted them to all men, as every man had need." They have maintained this extreme form of comprehensive cooperation throughout the centuries, in the face of trials and tribulations, persecution and martyrdom, and in spite of forced migrations from one country of Europe to another, and from Europe to the U.S.A., where they landed in 1874. Here they settled in South Dakota, mainly because this sparsely populated state allowed them to put a safe distance between themselves and the other inhabitants.

It may be worth noting that the few historic facts just mentioned definitely disprove *Amana's* claim to being the oldest religious cooperative community in existence. This claim can be upheld only as far as residence in the United States is concerned. The prolonged existence and steady growth of the Hutterite colonies equally disposes of the contention that "cooperation cannot work." It certainly did and does work with the Hutterites.

We knew that the basis of their organization was religious communism; that the Preacher was the highest authority among them; but that in all his decisions he had to consider the will of the General Assembly of all male adults; that each

chap. ii, Saul M. Katz, *The Security of Cooperative Farming* (Unpublished Master's Thesis), Cornell University.

member, including the Preacher, had to put in a good day's
work; that they adopted the most modern methods in farming;
but that they clung to their old-fashioned frugality in all ways
of life. But all this made us only the more anxious to meet
these people in person.

To do so, it would have been unwise to go straight to *Jamesville*. We had received due warning on this point from our
predecessors. The Hutterites are likely to be very aloof with
visitors not properly introduced. Their truly biblical hospitality
is reserved for those whom they know to be their friends. Such
a friend of theirs was Mrs. Nana Goodhope of Viborg, South
Dakota, who had kindly served as a guide to our research
men. Mrs. Goodhope had offered the same service to us.

We were warmly received by Mrs. Goodhope, treated to a
tasty lunch, and to the story of our hostess' friendship with the
Hutterites. It had begun during the First World War. The Hutterites, true to their faith, refused military service. Their habits
were strange, their dress foreign, and their language, the Ty-
rolese, was a German dialect. It was easy to interpret their
refusal as sympathy with the enemy. This interpretation was
the more palatable as the colonies had become quite prosperous. An opportunity to display one's patriotic zeal and to
enrich oneself in the process seemed to offer itself, and some
of the neighbors were unwilling to let it pass. To create the
right mood for their maneuver these worthy "patriots" smug-
gled broken glass into the Hutterite flour-mills and saw to it
that this "dastardly attempt upon the lives of loyal Americans"
was dramatically discovered. Thereupon several of the Hutterites were arrested. It was when two of them, quite young
yet, died in prison, that Mrs. Goodhope's ire was aroused. She
took up the unpopular case of the Hutterites and aired, in
several articles, her indignation about the injustice done them.
In spite of her efforts on their behalf, the Hutterites had to

abandon their villages and to take up, once more, their centuries' old trek, this time to Canada. It is here that the majority of all the Hutteritte colonies is now located. When the war was over and hostility was turned to guilt-feelings, the Hutterites were urged to return to their villages. *Jamesville,* which we went to see that afternoon is one of these restored villages.

Unlike *Amana, Jamesville* lies at quite a distance from any paved highway; we covered twenty miles of gravel road to reach it. On the way there, we learned from Mrs. Goodhope that things had been different during the Second World War. This time, the Hutterites' conscientious objection found official recognition. As farmers, they were classified as indispensable and made exempt from military service. Curiously enough, this sensible treatment resulted in another kind of trouble for the Hutterites. Two of their young men—and the sons of the Head-Boss, or Business Manager of *Jamesville* at that—volunteered and were still serving in the Navy at the time of our visit. The Brethren would have preferred persecution to this betrayal of the basic tenet of their faith.

We drove into *Jamesville* watching for signs of the unusual. But all we saw at first, was a gas pump, and the upper story of some large buildings. Mrs. Goodhope explained that these structures, executed in the functional style of plain Government architecture, had been added when the abandoned village was taken over and run for a short time by the Rural Administration. The picture changed quickly while we proceeded. Here were the bearded men with their wide-brimmed hats, long dark-grey or black coats and loose trousers; here the women, with black or polka-dotted kerchiefs, black full skirts, and long-sleeved waists reaching to the neck; and the children, in clothes that repeat the somber dress pattern of the parents, eyeing us strangers shyly. We drove into a wide yard, fringed on one side by what were apparently the utility buildings, and on the other by two rows of long stretched two-story structures. These were the buildings we had noticed before; they were the living quarters of the community.

Our coming had been announced, and as soon as our car stopped, we were approached by an elderly Hutterite woman, the wife of Joseph Waldner, the Head-Boss of *Jamesville* and three grown-up girls. She greeted us warmly and was very pleased when she found that she could talk German to us. She introduced the girls to us, her daughters, Barbara, Rachel, and Sara. The three girls took over the conduct of our tour through the community. First, we had to see their own apartment. The living quarters proved to be by far more comfortable than their outward appearance indicated. Each family, they explained, had as many rooms as it needed. They themselves lived in four rooms, two for the girls upstairs, and the living- and bed-room for the parents downstairs. There were no kitchens in the apartments: all the food for the whole community was prepared in a central kitchen and served in the common dining hall. There were also no other of the "conveniences" to be seen; there were no bathrooms, and instead of flush toilets there were outhouses; but there was electric light. To reach the upper floor, we had to climb an outside stairway which ended in a landing protected by a wooden railing. All furniture was solid, old-fashioned, and dark. The only colorful pieces were the girls' hope-chests of blond maple, decorated with gay and many-colored ornaments. The upper rooms were bright and spacious, and there was nothing cloister-like about them. Neither was there anything nunlike about the girls, of whom at least one, Sara, the youngest, would have passed anywhere as a beauty. The girls were not at all shy about showing us their trousseaus. While doing so, they joked about sweethearts, and kept teasing each other gaily, just like any other girls would. They spoke freely about their plans to marry and have families. They all thought they would prefer to marry into one of the Canadian colonies. They themselves had grown up at McLeod, in Alberta, and had come to *Jamesville* only eight years ago. They were proud about the position their father had achieved in the new place in such a short time, but they wanted very much to go back to Canada, mainly, it seemed, because it was cooler there.

We took photos of the girls; they posed willingly before their hope-chests and before the spinning-wheel which, by the way, is by no means just a piece of decoration here: all the wool used by the Hutterites is home-spun.

We were then taken on a tour through the village. We saw the central kitchen, with its whitewashed walls and tremendous coal stove, and the adjoining dining hall, a large low-ceilinged room with two sets of tables and benches, one for the men and the other for the women. We had a look at the stables, the bakery, the shoe and the broom shop. Everywhere we went, we found the Brethren and the Sisters tending their jobs earnestly, but without haste or compulsion. Things were kept in order, with no attempt at rigid neatness such as we had found at *Amana*. There was about the whole place and its people an atmosphere of peaceful satisfaction.

When we returned to the Waldner home, we found the table in the living room set for us. Although all eat in the common dining hall, the head-officers have the right to serve guests in their own home. Mrs. Waldner insisted that we have dinner, and we sat down to a meal of pickled ham and beef, salad, fried potatoes, dark tasty bread, and a bottle of dark-red, sweet wine. While we were still at it, Joseph Waldner came in, a tall, broad shouldered man of about fifty, with a curly graying beard, and a pair of clear blue eyes whose expression could probably best be described as "with malice toward none." We shook hands, and felt immediately at home with each other.

Later, while walking over to the barnyard to watch the girls milk the cows in the open, we had become good enough friends to speak about the affairs of the colony. Although *Jamesville* is the smallest of the four colonies in South Dakota, and owns only 2,738 acres of land as compared with 7,200 acres of *Bon Homme,* it does well. They live comfortably and know no want. At the end of each year there is usually some money to put away. They have no need for luxuries, and money above what they need does not change their habits. All they ever use it for is to buy a piece of machinery that makes work easier, or to give the young ones a chance to start out on their own. They do not permit any colony to grow larger than 150 or 200 souls, depending on available land, and establish, rather, daughter colonies.

Our talk turned naturally to Brother Waldner's own family. Of his twelve children only the three girls remained unmarried; the others had all established their own families, at *Jamesville* or in other colonies. And two sons were in the Navy—he mentioned it without any noticeable reluctance, but sadly. Their action had been a hard blow for him. "But," he pointed out, "we believe that everybody is responsible to God alone. They have done what they believed was right, and we have to trust that they will find their way home again."

At about seven o'clock, a boy began to make the round of the village, calling the faithful to the prayer-meeting. We followed Brother Waldner to the school-house which also serves as church. He explained to us that only the wealthier of the colonies could afford a separate church building. Attendance was not compulsory at these evening meetings, and the room was only half filled; as in the dining hall, men sat apart from the women. No sacred paraphernalia were visible. Essentially, the service was a communal hymn singing affair, with some reading from the *Grosse Geschichtsbuch,* the "Bible" of the Hutterites, in between the hymns. The singing was led by the Preacher—also Head-Gardener—a short, stocky man, with a brown beard, much younger than Brother Waldner.

The Preacher is the one to watch the morals of the community. We were given a sample of his powers in this respect when we took leave from the Waldners. They all, the parents and the daughters, stood around us, and we felt loath to part. Sensing this, Mrs. Goodhope invited Sara to come with us to Viborg, and to spend the night in her house. Father and mother readily agreed. But not so the Preacher, who was with us. He declared: "The best place is home," and the matter was settled.

Bon Homme, which we visited the following day, is much wealthier than *Jamesville;* but there is virtually no difference between the two colonies in their standard of life. The same plain clothes, the same dormitorylike living quarters, the same kind of home-grown food. Our guide at *Bon Homme* was Michael Waldner, the Preacher. With his short white beard and his pink complexion he looked like St. Peter himself. At sixty-eight, he was still doing manual work in addition to his preaching; he was in charge of the colony's timber. *Bon Homme*, which is the oldest colony in the United States, owns one of the most complete sets of Hutterite writings. These are Michael Waldner's pride, and we had to admire them. He told us that he had added a few old Bibles to the original collection when, in 1937, he went to visit Eberhard Arnold's *Bruderhof*, which was then still located near Fulda, in Germany. We were surprised to learn that a Hutterite would ever travel so far. He admitted that trips abroad were isolated instances among them; but though they tended to keep to themselves, quite a bit of traveling went on between the colonies, both in and between the U. S. and Canada.

The impression we had gained from our visit to *Jamesville* was heightened by our short stay at *Bon Homme*. We left the

Hutterites with the feeling that we had seen a people who had
found the secret of material and spiritual security. Watching
them, it all looked so simple: each of them worked steadily
at the assigned job, and tried to do it as well as he knew how;
when they earned more than was necessary to satisfy their
wants, they put it away and used it only to improve their
work, or to give their offspring a fair start. They were gay
and satisfied; they were obviously not scared of life, and death
could not hold much fear for them. If any human beings deserved that name, they certainly had a claim to being called
happy.

The Quest for Security

We were struck by the realization that we had left all the
other groups at best with a question on our mind. Cooperative
effort had secured for the coal miners of *Penn-Craft* better
housing; but would they learn from their success and go on
from there to secure for themselves all the benefits of cooperation? The searching minds of the boys and girls of the Yellow
Springs group had reasoned out well enough the advantages
of cooperative community in our atomic age; but was reason
alone potent enough to carry them—through the perils inherent in a departure from accepted ways of life—into cooperative living? To say nothing about the *Amana* people who, as
if frightened by the competitive uproar around them, had
abandoned in panic their old, cooperative tradition; they were
making more money, but were they better off now? The only
ones who seemed to have the answer were the Hutterites.
They had, long ago, crossed the border that divides competitive from cooperative living, and were satisfied to stay there
for all days to come.

The Hutterites' answer appeared to be simple enough, we
had found. It was only that once one began to think about it,

one felt like contemplating a treatise on cooperative as compared with competitive society. A treatise which would have to touch upon such fundamental contrasts as production for use and production for profit; how the one takes care of the needs of all, while the other provides plenty for a few and precarious subsistence for the many. It would have to show how the one husbands its resources and endows all with the sense of economic and psychological security; while the other disposes of its surplus in the wasteful way of "conspicuous consumption," and keeps even its most favored members in a state of never wholly relieved anxiety. Further in the text, this treatise would probably have to face the argument that no progress was possible without incentive, and that profit was the only effective incentive for production. Which argument could be easily disposed of by reference to the fact that it was by no means the inventive but rather the exploitative mind that was stimulated by profit only: inventiveness was much less hampered in cooperative than in competitive society. This would bring up the problems of efficiency and the dangers of shirking. Here, the findings of industrial research would be of use; they show that, above the subsistence level, increase of output is far less affected by wages than by *morale*. Or better still, the Israeli *Kvutza* could be cited as proof that rather than shirking it is "exaggerated devotion to work" which may cause embarrassment in the cooperative community. Finally, our treatise would have to point out that it was in the cooperative society that all were equally well housed, fed, and clad, and shared equally in all provisions for health, education, and recreation; while in competitive society, however prosperous it may be, there were always to be found those seemingly ineradicable evils: dependency, malnutrition, slums, delinquency, and crime.

But since there is no room here for such treatise, we should like to round off this report of our journey by one final consideration. It may be that the crux of the whole issue is simply good husbandry. Whatever the particular circumstances, anybody will easily feel secure when he is able to earn more than he feels like spending. It is only that needs, above mere subsistence, are conditioned by a given cultural setting. The mores of competitive economy tend to foster exhaustive spending. When status becomes dependent on "keeping up with the Jones's," only those in the very highest income brackets are able to enjoy a comfortable margin between what they earn and what they are forced to feel like spending. It is rather where the mores disparage outdoing others in waste and extol instead concern for the well-being of others that husbandry can become honorific. Where this happens, the sense of security ceases to be a privilege of the few and becomes common to all. A culture of this kind can be found today in cooperative communities.

On our trip we had encountered only one full-fledged instance of such community: the *Hutterische Gemein'*. If a continuous existence of more than four hundred years meant anything, there was the proof that this kind of socio-economic organization had the power of survival. But the community of the Hutterites had remained confined to observants of a sectarian creed, relatively insignificant in numbers. It could be termed "utopian" and dismissed as a freak in the sociological side show.

Yet, we were about to cross the border into a province of Canada where recently farmers had begun to take definitive steps towards the establishment of cooperative farms. Their chief motive was, unlike that of the Hutterites, not religious but rather that of the Rochdale weavers: economic self-help. It was the same motive that had brought forth the *modern* cooperative community, the Israeli *kvutza*, the Soviet *kolkhoz*, and the Mexican collective *ejido*. The Saskatchewan farmer could, and did learn from these foreign examples; but to make their experience effective he had to adapt it to his own cultural pattern—which is virtually the same as that of the American farmer. The cooperative community *he* developed could be dismissed neither as "utopian" nor as foreign to our own cultural setting. Was there an unexceptionable answer being designed to the quest for security across the border?

(5)

SASKATCHEWAN:
A PATTERN OF SOUND GROWTH

A Spontaneous Development

Cooperative settlements are found today in many parts of the world. Their number is considerable, and it may well be said that they form a substantial part of our modern socioeconomic reality. Social research should find in cooperative communities rewarding objects of investigation, for many reasons, but above all, because: *(a)* they function along clearly-defined lines of action, as laid down in the Rochdale principles; *(b)* the basic conditions of these communities are the same as those determining any human society; *(c)* being simple, but at the same time comprehensive social systems, they render the larger social whole more intelligible; *(d)* they can be subjected to a thorough investigation on the sociological as well as on the economic level.

There is, moreover, one noteworthy aspect of the cooperative community. Each new group as it forms reveals some of the mechanisms which must have been at work in the origins of civilized society. Insight, otherwise hardly accessible, offers itself to the sociologist into cooperative group formation.

It was with great interest, therefore, that the Rural Settlement Institute, now the Group Farming Research Institute, Inc., received the news of a movement towards cooperative farming among the farmers of Saskatchewan. The movement appeared to be wholly spontaneous, aided by genuine sympathy and support on the part of the provincial government. Several farms, we learned, had been already incorporated, and others were in the process of becoming established. No large numbers were involved. But the fact that a sound and

94

promising trend towards the modern cooperative community seemed to be developing in a cultural pattern so similar to, if not identical with that of the U.S.A., was enough to warrant attention. It was decided to look more closely into this development and during the summer of 1946 a field survey was carried out which took the author, with a small supporting staff, through all the incorporated cooperative farms in the province. The number of these farms has since more than doubled and was 27 by December 1952. They were all organized in a way identical or quite similar to the farms covered in the following survey.

The Need for Cooperative Farming

A basic assumption in all systematic investigation of cooperative groups could be formulated as follows: Between certain basic needs and their satisfaction, our competitive culture has erected barriers, which are experienced by an increasing number of individuals as insurmountable. Instead of resigning themselves to frustration, some abandon, to a lesser or larger degree, competition, with its emphasis on doing things against others for one's own sake, and resort to cooperation. In doing so, they are ready to yield voluntarily part of their freedom and to share the benefits accrueing from their common enterprise equitably.

If this basic assumption is correct, the first questions to be answered are these: *(a)* what are the particular needs which cooperation tries to satisfy? *(b)* what are the particular barriers in the given situation which cooperation tries to overcome? Applying these questions to the situation in Saskatchewan, we shall ask: What are the farmers' needs in this province left unsatisfied by the competitive economy? And what aspects of this economy make them resort to cooperative action?

The business of the Saskatchewan farmer is essentially the

same as that of farmers anywhere. It consists in the endeavor to derive from the cultivation of the soil and from related activities a reasonably secure livelihood. In his particular case, there are three basic needs which it is difficult for him to satisfy; (1) economic security; (2) adequate social contacts; (3) fair provision for the next generation. In addition, there is the need, created by the termination of the war, to provide veterans desiring to take up farming an opportunity to do so.

Due to geographic-climatic and soil conditions, the Saskatchewan farmer appears to have been forced into the most competitive kind of agriculture, one-crop farming. The whole southern part of the province is particularly well suited for wheat farming. As long as prices are good, such farming proves quite profitable; but as soon as prices fall, the situation can easily become catastrophic. The story of the years from 1920 up to the present is one in which "the general level of prices for Canadian farm products probably experienced the most violent fluctuations ever recorded in the economic history of Canada." [1] Prices fell from an index of 164 in July, 1920, to 96 in December, 1921. After a gradual recovery until 1925-26, prices fell again until 1929, when the world-wide depression made them hit rock-bottom. While things were bad in Canada as a whole, they were even worse in Saskatchewan. At a time when the lowest index for farm products for Canada in general stood at 48.4, in Saskatchewan it reached a low of 33.9.

In general, this was due to the drastic decline in wheat prices, but in particular to what Hope and his associates call "the rigidity of general distribution charges." [2] Which, in other words, means that the railroads, exercising their monopoly of

[1] E. C. Hope, H. Van Vliet and C. C. Spence: *Changes in Farm Income and Indebtedness in Saskatchewan during the Period* 1929 *to* 1940. Univ. of Saskatoon, College of Agr., Extension Bull. No. 105, p. 6.
[2] *Ibid.*

transportation, kept exacting the same rates they had been charging when the price of wheat was three times as high. The same "rigidity" of the competitive system held living costs and the price of "things farmers buy," out of step with the falling income. While the index of farm prices reached the low of 33.9, the index of living costs remained up at 80.9 and that of "things farmers buy" at 82.8.[3] No wonder that when, into the bargain, the grain fields showed a low yield, as happened in Saskatchewan during the period from 1929-40, the net income fell to such low levels as to drive many farmers into bankruptcy.

Relief from this situation was sought in cooperative action of the segmental kind. The history of the wheat pool in the last two decades is the story of a successful fight for a more stable price for farm products.[4] But there were many whom even price stabilization could not help. Those "located on lands marginal and sub-marginal for wheat production, and some of those farming in districts ordinarily considered to be fair farming areas, came to the conclusion that farming under these conditions was a hopeless task." [5] Some of these defeated farmers emigrated to the U.S. or elsewhere; some went north, into the wooded area of Saskatchewan. Here the land was cheap and could be obtained by homesteading. But the difficulties were enormous. The land had to be cleared and broken, and this could not be done fast enough to provide even mere subsistence. Relief and re-establishment aid were needed to save the resettled from starvation, long after farmers in the wheat region had begun to recover. Only during the last years

[3] cf. 1. cit., Table I.

[4] See *The Wheat Pool and its Accomplishments,* issued by Saskatchewan Co-operative Producers, Ltd., 1946.

[5] R. A. Stutt and H. Van Vliet: *An Economic Study of Land Settlement in Representative Pioneer Areas of Northern Saskatchewan,* Dominion of Canada, Dept. of Agric. Technical Bull., No. 52 (June 1945) p. 7.

of the war, with high prices for farm products, could these northern settlers reach self-support. Only then action aiming at self-help became conceivable.

The main need was more efficient clearing of the bush and breaking of the land. Machines, which the individual farmer could not afford, were needed, and so the first step towards self-help assumed naturally the form of a machine-sharing cooperative.

In Saskatchewan, most of the cooperative farms are located in the northern parts of the province. Pooling of machinery is the main feature; but it is often only the initial step towards a more inclusive kind of cooperation. Four of the incorporated cooperative farms were at the time of the survey of this type: *The Round Hill* and the *Mount Hope Agricultural Production Cooperatives,* in the northwest; the *Algrove Cooperative Farm* and the *Orley Production Cooperative Association,* in the northeast.

Another need which cannot be satisfied on individual farms in Saskatchewan is that of adequate social contacts. In the gamble of one-crop wheat farming, only the successful can escape the dreariness of the isolated farm. They do so, by the simple device of closing up for the winter and moving to a home in the city. For the less well off, as soon as the roads are turned into "gumbo" traps by rain, or into impassable drifts by snow, there is complete seclusion. Even when the roads are open, distances between the individual farms are so great as to make social intercourse practically impossible. With all that they imply in lack of technological, medical, educational, and recreational services, the unsatisfied social needs prove thus an additional factor in the trend towards the cooperative farm. This factor was the dominating motive in the formation of the *Sturgis Farm Cooperative,* located in the north-east of the province.

A third basic need is that of providing for the young on the farms. One-crop wheat farming requires a minimum of a quarter section (160 acres) for a bare subsistence. Partition of a small, or even a middle-sized farm among the male off-spring means condemning each of them to failure from the start. The oldest son alone can hope to take over his father's farm and to make a go of it. If there are younger sons, they have no choice but to hire themselves out as farm hands. They may hope to save, after years of strenuous effort, enough money for a down-payment on a small farm of their own. There are some who succeed in doing it; but they labor under a heavy debt, and only a few are fortunate enough ever to obtain a clear title to their land.

One family took the initiative in showing a way out of this unfair routine. The farm-owning members decided to pool land and equipment, and to make the younger sons full-fledged members of the cooperative farm. On the *Laurel Farm,* located north of Saskatoon, the younger sons start out as co-owners and work towards full equity in the cooperative. Since then several such family farms have been founded.

Finally, there is the problem of the boys who went into the Army. After they helped to win the war, they expected an opportunity for a better, more secure, and socially more satis-fying life on the farm. Many of them, if they were married, had a reasonable chance of obtaining, with the financial help of the Federal Government, an individual farm. 1,200 veter-ans were settled in this way in Saskatchewan by 1946. But this was practically impossible for the single veteran, and for this reason the Saskatchewan Government decided to offer settlement to war veterans in cooperative farms. The response to this offer was strong enough to warrant establishment of the Veterans' Cooperative Farm at Matador, in the south-west of the province. Its success has since led to the establishment of 11 more farms, of this kind.

It should be understood that by singling out the dominant

need which cooperation was in each case found to be satisfying, and by thus formulating four basic types of cooperative farms in Saskatchewan, we do not mean to imply that these motives function independently of each other. As a matter of fact, in each case, all the needs enumerated will be found to supply a motive for cooperative action. It is only for the sake of greater clarity that this method of presentation was chosen. And also, in the hope that it will help others to identify their own needs and to profit from the manner in which the Saskatchewan farmers went about satisfying them through cooperation.

The Economic Needs

At the time of the survey the Department of Cooperation listed 32 cooperatives for agricultural production in Saskatchewan. Among these were nine cooperatives for the use of machinery and cooperative farms, "progressing toward cooperative communities."

The oldest among the machine sharing cooperatives in Saskatchewan is the *Round Hill Agricultural Production Cooperative Association,* incorporated in November, 1943. It is the only one of those here described which was established before the present Government came to power in the province.

Machine sharing of the more or less informal kind has been practiced among the farmers of Saskatchewan prior to the incorporation of *Round Hill.* But wherever it was attempted with a more extensive scope it became ensnared in serious difficulties. Chief among these was the problem of organizing the use of machinery to the satisfaction of all concerned. Farmers in the same area will often want to use the same piece of machinery at the same time. Those whose turn comes too late will resent the bad deal and will be inclined to give up the arrangement. The *Round Hill* group appears to have found a satisfactory way out of these difficulties.

Farming in the North Battleford area, where the *Round Hill* group was formed is of a diversified character, crop growing combined with livestock breeding. The average farmer cannot handle efficiently both branches of agriculture simultaneously. He has to sacrifice a large part of his income to hire a man, or his farm will suffer. The way out of this dilemma was found in a cooperative which pools the machinery, hires one man for the whole group to handle it, and frees the members to attend to their livestock.

A driving force behind this cooperative was a couple with an unusual background. The man had been a *Junker* of the purest calibre, officer in the German Imperial Army during the first World War, heir to a large estate in East Prussia, and owner of Silesian coal mines; his wife, the daughter of a German aristocratic family with close relations to the Imperial House. The inflation of 1923 ruined them financially and drove them to emigration. After several years of attempting to re-establish themselves in the U.S.A., they landed in Canada and went into farming, first hiring themselves out, and later becoming tenants on their present farm. Thus re-conditioned to the ways of democracy, they took the lead in educating their neighbors in the principles of cooperation and in November, 1943, eight of them registered the *Round Hill Agricultural Production Cooperative Association, Ltd.* Two more joined a year later, and the membership has remained since then at ten.

The stated purpose of the association was, "To purchase, rent, lease or otherwise acquire farm machinery and equipment to be maintained and used by the association to assist its members to carry on their farming operations." [6] They elected a board of directors, as required by the Cooperative Associations Act, and proceeded to operate. A manager was elected

[6] Quoted from an unpublished Master-Thesis by Leslie E. Drayton (University of Alberta, 1947).

and a man was hired to operate and maintain the equipment. The manager was given the power of important decisions. It was up to him to determine, after consultation with the directors, "the charge to be levied on each member for the use of machinery and equipment in performing various operations," such charge to include: *(a)* the basic cost rate for the use of machinery on an hourly basis; *(b)* local labor charges per hour, including local fuel and oil costs; and *(c)* charges for transportation between farms, where this appeared warranted. All these charges were to be paid promptly, a member in arrears in payment for more than two months to be excluded from further use of the association's equipment.

It was provided that equipment in possession of members could be hired or purchased by the association at the discretion of the manager, who might also employ individual members in the interest of the association, as he saw fit. Compensation for work done by a member was to be credited against his debts to the cooperative for services received. The manager was also to be the final judge as to the efficiency of work performed by any member. Appeal against his decision could be made to a special arbitration board, appointed by the directors.

As to the order in which members obtain use of the various machines, the power of decision rests also with the manager, while the directors function in a consultative capacity. Each member is, of course, assured of all due consideration, but the manager's decision as to where and when each machine is to be used, is final, and no appeal can be made against it. Only in renting the machinery to non-members is the manager dependent on the consent of the directors.

The manager has no say in the acceptance or retirement of members. Such matters are reserved for the directors and for the meeting of members. Retirement of a member can be decided upon when the directors find that he is not using the equipment in a "fair and reasonable manner," or if he persists in delaying the payment of his accounts, or if he otherwise fails in the performance of his duties and obligations. The directors' order of retirement must be confirmed by a two-thirds vote of a membership meeting. The retired member's "outstanding account, if any, may be deducted from his share capital and his loan capital, if any, and the balance refunded to him."

Guided by these arrangements, the *Round Hill* group has done quite well. Having run their farms with horses prior to forming the cooperative, they purchased, in 1944, a tractor, a tiller combine, a cultivator, a grain crusher, and a separator, at a total cost of $3,835. Members were employed by the cooperative for threshing only, and did their own seeding, cutting, and stooking. The surplus of the first year amounted to $248, in addition to $857 allowed for depreciation, which sum probably contains part of an undivulged surplus. After setting aside small reserves, the surplus, it was decided, was to be divided among the members in proportion to the value of services they received from the association.

In the following year, more machinery was acquired, a tandem disc, a disc packer, an oil wagon, and a 28-run drill. The association now owns a full line of modern equipment at an average investment per member of $500, and operates a total of 1,400 acres, composed of nine individual member units.

The *Mount Hope Agricultural Cooperative Association,* located in the same area, was formed two years later than *Round Hill,* in 1945, and has obviously profited from the experience of the older cooperative. Being somewhat larger—it has 15 members and operates a total of about 3,500 acres—it is organized along the same lines as *Round Hill.*

Both cooperatives now follow the practice of rotating the use of machinery by starting operations each year at a different point. In both cooperatives some of the members contemplate seriously further pooling of resources, particularly of land, and are making plans for developing their cooperative into a full-fledged cooperative community.

In a certain sense, this step was taken by the third group which still operates chiefly as a machine sharing cooperative. The *Algrove Farm Cooperative* owes its origin to slightly different motives. The members are farmers who were "dusted out" in the early 'thirties from the southern parts of the province. They obtained their quarter section in the bushland area south of Tisdale mostly by homesteading. None of them succeeded in clearing more than 40 acres for cultivation and many of them were on relief until a few years ago.

There is something desperate in the resolution with which these men go about helping themselves. Their cooperation is born out of the fear that "prosperity" can not endure, and out of determination to keep clear of relief. Their first need is more cultivated acreage on their farms. To speed the clearing of the land from the bush, heavy machinery is required. So, after studying the problem for more than a year, sixteen of them got together in March, 1946, and organized their cooperative. Pooling their meager resources, they raised the sum of $1,800. This was not enough to purchase the necessary equipment. They were aided by a Government guarantee in securing from the Saskatchewan Cooperative Credit Society a loan of $3,600. With the money thus raised they bought two tractors and some other small equipment. One of the tractors is used to cut the bush and the other to break the land; 540 acres were brought under cultivation during the first season. The group intends to allocate surplus on the basis of labor contributed, use of machinery, and purchase of produce from the association. But they have decided to delay distribution of dividends, and to deduct a certain percentage of the wages in order to build up their reserves.

The plans of the *Algrove* group go beyond mere machine

sharing. They have already leased cooperatively three sections from the provincial government; 250 acres of this land have been cleared and broken during this summer, and will be seeded next year to supply feed for the members' livestock. Later on, they would like to pool all their capital resources, move their homes to an already selected site, and to form a full-sized cooperative community.

It may be expected that the *Algrove* people will eventually succeed in carrying out their plans. They seem to have the necessary persistence, and they have had the good fortune to find efficient leadership. How much both are needed to make headway, may be seen from the experience of another cooperative, formed in the same area a year earlier. The *Orley Production Cooperative Association*, formed by about 30 men in 1945 for the same purposes as *Algrove*, suffered a crop failure in the first year of its existence and was unable to raise the sums necessary for the purchase of heavy equipment.

The Social Needs

The most advanced and the most significant of the groups here discussed is the *Sturgis Farm Cooperative Association, Ltd.* This for several reasons, of which we should like to mention the following:—

First, the members of this group were quite successful as individual farmers. In resorting to cooperation they should, therefore, have the best chance of succeeding. For, experience shows that an important factor in cooperative community success is a membership which does not seek in cooperation escape from personal maladjustment. This does not mean that failure under a competitive set-up lessens *a priori* the chances for success in a cooperative. Such chances will depend rather on whether failure was due to the iniquities of the system, or to personal shortcomings. To determine this in each case is difficult, if not impossible. Inquiries of this kind will not be urgent with people who were able "to make a go of it" under competition. They will see in cooperation more than a mere economic convenience. Their "uneasiness" will have its roots, not so much in economic insufficiency, as in the need for a more satisfactory society.

The nine charter members of *Sturgis* did not enter their cooperative as poor men. Each family was able to loan to the association about $8,000, in addition to paying $100 as a membership fee per person. The *Sturgis* cooperative began activities with a loan capital of $43,627.

Second, in deciding on an all-out cooperative community from the very start, the members must have been aware of some of the crucial problems of the cooperative movement. The nature of these problems was first revealed by George Russell, in the famous chapter VI of his *The National Being*. "It is not enough—Russell says—to organize farmers in a district for one purpose only—in a credit society, a dairy society, a fruit society, a bacon factory, or in a cooperative store. All these may be and must be beginnings; but if they do not develop and absorb all rural business into their organization they will have little effect on character. . . . The specialized society only develops economic efficiency. The evolution of humanity beyond its present level depends absolutely on its power to unite and create true social organisms." [7] Reference to these problems was made again later in the critical analysis of the British cooperative movement, by Carr-Saunders *et al.*[8] After pointing out that "the influence of the movement, despite its spectacular success in the sphere of trade, has been negligible in the realm of ideas," they state that this defect was "closely connected with the failure to work out a

[7] A. E. (George Russell): *The National Being* (New York: Macmillan, 1930), p. 40 f.

[8] Carr-Saunders, Florence and Peers: *Consumers' Co-operation in Great Britain* (New York: Harpers, 1938), p. 47.

new philosophy of cooperation after the older Owenite ideal had been abandoned."

The assumption that the "true social organism" of the modern cooperative community might well turn out to be Owen's "village of cooperation" seen in the light of "a new philosophy of cooperation" must certainly have been in back of the minds of the *Sturgis* people when they made their all-out decision. Most of them had been members and officers in "specialized" cooperatives, the wheat pool, marketing and purchasing cooperatives, etc. But they felt that this type of cooperation was not sufficient to relieve their "uneasiness in the competitive culture." They formed a study-action group and after considering their problems for two full years, they concluded that the answer was the cooperative community. Their enterprise thus carries implications that may prove important beyond the limits of their immediate achievements.

Third, being leaders in the wider community—one of them is an M.P., another a delegate to the county council—the members enjoyed a high social status before joining the cooperative. This is worth noting, because of the suspicion, or even contempt, with which Anglo-Saxon populations are inclined to view any such radical departures from the established norm. One of the severest handicaps under which the Farm Security Administration's cooperative farms in the U.S.A. had to labor was the distrust of the general public. Formed by highly respected citizens, and not by people on relief, the Sturgis cooperative farm does not meet with any such prejudice.

There is yet another hazard in such ventures which should be lessened by the high social status of those who embark upon it. Even the successful cooperative community has in the past rarely been able to prevent the breaking away of the next generation. The younger people resented being thought of as "different" by the outside world. In the case of *Amana* this resentment was one of the main reasons for the shift to a joint stock company, and in New Llano the young tended to quit the community long before it collapsed. Given the high social recognition, it may be expected that at *Sturgis* the children instead of disavowing their parents will be proud to follow in their footsteps.

Finally, there is the high educational level of the members. Three of the men and two of the women have a university background; one of the men has been instructor at the University of Sackatchewan. These men were able to approach their problem analytically. Once they had mapped out their plan of procedure, it proved to be of a calibre which made the supplemental by-laws of *Sturgis* a model for all cooperative farms incorporated since in Saskatchewan.

The charter members of *Sturgis* had been farmers with fair-sized property of their own. But to make requirements for joining truly cooperative, they decided that: "Any persons, either owners or tenants of land, or any persons who may contribute their personal services to the association," were eligible for membership.

Whether property owners or not, all members had to pledge themselves to "contribute whatever services in the interests of the association may be required from time to time by the manager and the board of directors." One who owns property, has to "describe and indicate the amount of capital and resources" he is prepared "to subscribe and contribute to the assets of the association." Since, in the case of farmers, such capital and resources consist chiefly in values which are subject to change, the applicant has to agree "to any special revaluation of assets or capital and resources of the association from time to time as may be determined by the board of directors." To guarantee the utmost fairness to the member in reference to such revaluation, he is given the right of extensive appeal. He may first turn to a meeting of the members or redress; if the decision of the meeting does not satisfy him, he is free to seek arbitration by means of a tribunal "which shall be composed of a representative appointed by the member, a representative appointed by the board of directors together with a non-member, approved by the Registrar of Cooperative Associations, in the event that the other members cannot agree as to who shall be the independent member."

The applicant has further to agree "to empower the association to retain any loan made by the member, or his membership fee, for a period of time to be decided upon by the board of directors." But in case of retirement, full restitution of the "amounts held to his credit," including his membership fee is secured to the member, subject, of course, "to the terms of any special contractual arrangement which the member may have made with the association."

To those who admire the propertylessness of the Israeli *Kvutza,* such concern with property may seem to be at odds with the true meaning of comprehensive cooperation. Cooperation, however, is a technique of doing things. It should not apply dogmatically, but should suit itself to given conditions and needs. The membership of the *Kvutza* was recruited mostly from penniless youngsters who were wholly dependent on outside aid. The pattern they set has great merits. But it is impossible to apply it to countries like the U.S.A. or Canada without profound modifications. In this connection, the example set by the *Sturgis* by-laws, skillfully combining the spirit of cooperation with the requirements of fair business dealing, may prove of great value.

The *Sturgis* by-laws contain some other interesting provisions. In addition to more routine arrangements—such as defining rights and duties of the directors and of the managers (who is a member and is not allowed to hold office for more than two consecutive terms), they provide for payment of a "going wage" to each member, male or female. Since some of the members, like the M.P., serve with pay outside the group, they were confronted with the problem of how to handle such earnings. Here, their solution resembles that of the *Kvutza*: "All net earnings of members for services rendered to other than the association shall be paid into the treasury of the association, and such members shall receive standard wages during the period of time such services were given in the same manner as would have applied had they been working for the association. . . . If the outside work is of longer duration, the member may obtain leave of absence and then he may retain outside work without the consent of the manager." It is further provided that members who are called upon to perform services for the wider community, may obtain leave with full pay, for not more than one month a year.

The *Sturgis* group has been incorporated now for several seasons. Immediately upon pooling their resources which included land as well as equipment, they found that they could do without about a third of their machinery. They auctioned off the surplus for about $4,000. One of the farms in the pool was located at a distance of 25 miles from *Sturgis,* and required an extra tractor. They exchanged the farm for land close to the site of the community, and the tractor was sold. The group concluded even its first year with a substantial surplus.

On the 1,700 acres they cultivate, the group continued one-crop grain farming as practised before incorporation. They planned, however, to diversify their crops, and to add livestock as soon as they moved to the community site. The first steps toward building the community were taken in 1946. They purchased 80 acres of land near *Sturgis,* and laid the foundations for the new houses. The members, whose number has risen to twelve, helped in the building. The houses are owned by the cooperative. Rent is charged according to the size of investment.

Once the houses were built, cooperative community activities were taken up, step by step. As long as the community is not large enough to afford all such facilities as consumers store, church, school, community centre, electricity, etc., they utilize those of the nearby village of Sturgis. Later on, they plan to concentrate all social activities within their own community.

The Younger Sons

The *Sturgis* group has set, in its by-laws, a pattern which lends itself to emulation by other groups, even though their motives may differ from those of the original. The predominant motive in the organization of the *Laurel Farm Cooperative* Association, incorporated early in 1946, was the desire to treat the younger sons decently. Instead of resigning themselves to the prevailing practice of virtual disinheritance as a necessary evil, the nine adult members of the family, six men and three women, sat down together to think the matter over. After prolonged study, they found the way out in cooperation.

The older members of the family each owned enough good mixed farm land for a fair maintenance, the father 480 acres, the oldest son 160 acres, and the son-in-law as much as 1,500 acres. They had also modern equipment, and they had nothing to worry about as far as they themselves were concerned. But there were the younger sons who had nothing to look forward to but a never-ending struggle for mere subsistence, with hardly a chance of ever becoming independent.

The educational background of the group was, like that of the *Sturgis* people, quite remarkable. Three of the sons had finished agricultural courses at the University of Saskatchewan, and all three female members had been school teachers. Like the *Sturgis* group, they enjoyed the respect of the wider community. And they, too, realized that cooperative community, the pooling of all their resources, land as well as equipment, labor, and livestock, was the solution to their problem.

The *Sturgis* by-laws, with their provision for non-propertied as well as propertied applicants, lent themselves well to their purpose, and they accepted these by-laws in full. In practice, however, there are some differences, because of the different conditions. First, unlike *Sturgis*, the *Laurel Farm* started off with three farm owners and three members, the younger sons, who could contribute only their "personal services." This made for initial inequality on the economic side. They had to put, therefore, more emphasis on the provisions which aimed at remedying such unevenness, as: "In order that capital contributions from members, by way of membership fees, membership loans, and retained dividends, may be more equalized in amount, the directors may require a member who has an investment in capital contributions of less than 50 per cent of the largest member investment to allow any balance of dividends accruing to him for payment in any year to be placed to the credit of his membership loan account until such time as his investment equals or exceeds 50 per cent of the largest member investment." In other words, the members who offer only their labor to the pool let their share of the dividends stand until their equity approximates that of the other members. This may take a longer or a shorter time, depending on the good fortune of the group and the size of their surplus. But at no time does this impair the standing of the younger sons. They receive the same going wage, and they have an equal say in decisions. They differ only, for the time being, in terms of income. They have no claim to interest on capital, since they have not contributed any, and their dividends are retained to build up their equity in the farm.

The *Laurel* group has not, like the *Sturgis*, moved to a new site, but has set up its homes around that of the son-in-law. The eldest brother has literally picked up his home from a place about two miles away and put it up close by. The father, whose house was not worth moving, has built himself a new one next to the other two. The group decided that the members should own their houses individually. The sons, as long as they are not married, lodge with the parents. Should they establish families of their own, they will receive loans from the cooperative to build their own houses.

The nearest village, Meskanaw, is two miles from the site of the community. The group is therefore, more than *Sturgis,* dependent on its own facilities. They have their own electric power station, and they have already laid out a community garden and orchard. They also own fifteen cattle.

Finally, they have been able to realize, quicker even than *Sturgis,* some of the social advantages of cooperative farming. The members have already begun to specialize in the branches of agriculture for which they have most inclination; and some of them were able to afford, probably for the first time in the history of Canadian farming, the luxury of a three weeks' vacation with pay.

The *Laurel* group consisted originally only of members of the same family. But it was by no means their intention to re-main a family cooperative. They were willing to consider ap-plications by "outsiders," but they rightly hesitated to add any members before their own group had been fully consolidated. Recently several non-family members have joined the group, raising the membership to twelve.

The War Veterans

A distinctive variant, in the manner of formation as well as in motivation, is the *Matador Cooperative Farm Association.* This cooperative farm owes its origin to the desire of the pro-vincial government to offer the returning veteran, who wishes to settle on the land, a fair opportunity to do so. The members were not neighbors. They were brought together by the good offices of the government, and learned to know each other after settling on the land.

The plans for a veterans' cooperative farm go back to 1944. In August of that year at a conference called by his department, the Hon. J. H. Sturdy, Minister of Reconstruction, offered the following reasons in favour of veterans' cooperative farms: The men in the armed services received in this last war a sound training in handling machines; they also learned to appreciate the advantages of group action. It was not likely that, after their return to civilian life, men thus trained would want to go back to primitive methods of farming and to the dullness of the individual farm. Their best chance to utilize the acquired technical and social skills, to achieve a higher standard of living, and a more stimulating social life lay therefore in farming cooperatively.

To facilitate action, it appeared necessary to put through an amendment to the Veteran's Land Act. A resolution was passed by almost all cooperative and agricultural organizations in the province, and taken by the Hon. J. H. Sturdy to Ottawa. The resolution ran: "That the Veterans' Land Act be amended to provide that veterans may, if they so desire, pool the grants for which they may be eligible under the Veterans' Land Act to permit of their entering any properly constituted cooperative farming enterprise."

Next, a suitable location had to be selected for the proposed farm. Such a location was found on the *Matador* ranch, north of Swift Current, in the south-western part of the province. The provincial government owned sixteen sections (10,240 acres) of Crown land, which had been used originally by ranchers from the U.S.A., and later partly by Canadian ranchers and partly for communal grazing. A strip of land, three miles wide and seven miles long, was found to lend itself to large-scale farming, and the site was decided upon.

Finally, a call went out to men in the armed forces to apply for membership in the projected enterprise. Of the many who answered, the most interested were brought together in Regina, in April, 1946. For a whole week 26 applicants met for six hours a day to discuss with experts and government officials agriculture and cooperation in general, and the problems facing them at *Matador* in particular. As a result of the meetings, seven of the applicants dropped out; the rest had made up their minds to go ahead with the project.

Three weeks later all 19 had assembled at *Matador*. The arrangement under which they took up operations provided that, as long as the amendment was pending, they would receive wages, equipment, and facilities on loan from the Department of Reconstruction. Once their financial situation was clarified the veterans were to take over on their own, rent the land on a 33-year lease basis with option of purchase at the end of ten years, and to refund to the Government all sums advanced.

The conditions under which the group gathered were extremely difficult. For several weeks the men had to live in a caboose, at a barren place, five miles from the nearest road and six miles from the nearest farm. A first step towards decent housing was made when a large barrack of the St. Aldwyn's airport was cut into sections and moved to the site. One of the provisory buildings that went up served as dining hall, meeting center, and living quarters for a couple in charge of the kitchen, and the other, as dormitory for the men. Three crews were formed by the men, one to dismantle buildings at the airport; a second, to put them up at the permanent location of the community; and the third, to start breaking the land, with tractors leased from the Department of Reconstruction, and ploughs bought from farmers in the neighborhood.

By October, four houses and a dormitory had been erected; a barn had been bought and moved to the place; two bins had been built, and a machine shop was about to open; 2,500 acres

of land had been broken and worked down, 330 acres c
which had been seeded to flax. The number of members ha
fallen to fifteen, four of them married.

Twice in the meantime, the Hon. J. H. Sturdy had appeare
before a Parliamentary Committee in Ottawa to put throug
the amendment; both times without success. It had becom
clear, by August, that no decision could be expected in th
near future. In spite of the risk involved, the group decide
to incorporate. They accepted the *Sturgis* by-laws in full, an
elected a chairman, a secretary, and five directors; for lan
operations and field husbandry; for livestock and poultry; fo
landscape gardening, tree planting and irrigation; for mechan
ics; for construction; and for education and recreational acti
vities. A general meeting of the members is held regularl
each week.

So far as the work on the farm goes, the *Matador* grou
obviously made good, as good as, if not better than, the othe
cooperative farms. This should not be surprising in view o
the fact that all these men were born to farming. Before the
enlisted, most of them had worked for others, and had no
liked it. Now they had the chance to work for themselves, an
they thought nothing of putting in a few extra hours of worl
after dinner.

Really remarkable is the solid group structure which ha
formed within an unusually short period at *Matador*. It i
true that in contrast to the other groups there has been a rela
tively large turnover. Of the 19 men who started out togethe
at the end of April, four had dropped out by October. Bu
when it is recalled that the group was formed by men who
were total strangers to each other, and that the insecurity
about the V.L.A. grant was not conducive to group-integra
tion, a turnover of about 25 per cent, in the formative stage
must appear as not unduly large. A sociometric test, applied
in July with the purpose of establishing the stage of integra-

tion reached, showed that out of the 15 members participating in the test only one had remained an "isolate." The test revealed an unusually high degree of mutual attraction: they not only liked to work and live together, they teamed up also for outside activities, such as sport, hunting, and visits to the town. Already, after staying together for only a few months, their loyalty towards each other and towards their common understanding was quite pronounced. It expressed itself in such statements, as: "It is true that we have some squabbles, sometimes—but let any outsider dare say anything against any of us, we would stop him fast." Or: "I feel more at home here than I ever felt with my folks back home." And more challenging: "Let them try to get me out of here; they'd have to dig me out with a tractor."

This high degree of group-coherence, it may be assumed, has been made possible by similarities in the backgrounds of the members, and by the conditions under which they took up settlement. Of 14 members from whom data could be secured, eight were born in Saskatchewan, and four in other provinces of Canada; all 14 came from rural districts, having grown up or worked on farms prior to service in the armed forces; nine came from families with five or more children; while two had completed high-school and one had had some college training, six had completed not more than the eighth grade, and four had had one or two years of high-school; the youngest was 26 and the oldest 36; nine were Protestants, two Roman Catholics, and three of other faiths; ten were bachelors; and while seven gave economic reasons for joining, six professed to have done so because they were attracted by the "new way of life."

The plans provided for a combination of cooperative farming with individual living. But, as we have seen, conditions forced the men to take up from the start an extreme kind of cooperative living. No other arrangement was possible with only two houses available for 17 men and one woman. Without intending to do so, they had to emulate the *Kvutza* with its central kitchen, its common dining hall, and dormitory housing. It is probable that even more than similarities in the backgrounds it was these conditions which caused the rapid growth of a community spirit.

Justly encouraged by its first experience, the Department of Reconstruction has begun to clear land in the Carrot River valley for veterans' farms, and has succeeded by now in settling eleven more groups on cooperative farms.

The Political Climate

An essential characteristic of cooperation is voluntary association. Cooperation cannot be ordered, it must be freely chosen. Those who tried to disregard this truth in Russia paid with disappointment and failure. During one phase of "collectivization" over-zealous officials, eager to produce impressive and quick results, used threats and compulsion to force the *mujiks* into the *Kolkhozy*. The effects proved to be so disastrous that Stalin himself had to interfere.

Government, on the other hand, if sympathetic, can be quite helpful. The *Kvutza, e.g.* grew out of the spontaneous choice of the pioneers themselves. But without assistance of first the Palestine Office and later the Jewish Agency, the *Kvutza* would have never been able to achieve its amazing success.

The initiative for organizing cooperative farms in Saskatchewan came, as we have seen, from the farmers themselves. Even in the case of *Matador,* those who joined did so out of free choice. But while the Government wisely refrained from any interference with the budding cooperative farms, it freely offered every service at its disposal. In addition to the splendid "human material," it is this political climate which augurs so well for the future of cooperative farming in Saskatchewan.

Saskatchewan has always been in the forefront of the coperative movement in Canada. A chart indicative of the oncentration of cooperative marketing and purchasing asociations in the different provinces, drawn by the Dominion Economic Division, assigns Saskatchewan with 33 per cent he first place. Far behind come, second and third, Alberta with 16 per cent, and Ontario with 12 per cent. Saskatchewan, o quote an article in the *Saskatchewan News* (August 12, 946), "is where the word 'co-op' appears on grain elevators, milk wagons, oil trucks, honey jars, warehouses, funeral homes, and even tennis-court club rooms." To support the tatement, the article presents figures: ". . . about 70 per cent of the province's adult population belong to one or more coperatives. . . . Over 300,000 members (out of a total population of about 900,000) . . . with 1,000 co-ops operating 2,500 places of business in the province. In 1945 these coperatives computed their total assets at $71,000,000."

In February, 1944, in a pre-election speech, the present premier of Saskatchewan, the Hon. T. C. Douglas, declared: "Producers' and consumers' cooperatives have a basic part to play in a cooperative commonwealth. They would safeguard the interest of the farmer and the consumer in the same way that trade unions safeguard the welfare of the industrial worker." And he made this promise: "Under C.C.F. government, cooperatives would be encouraged and expanded." On June 75, 1944, the Cooperative Commonwealth Federation came to power in Saskatchewan, capturing an overwhelming majority of 47 out of 52 seats in the Legislature. On July 7, the new government took office, and on November 2 of the same year it established a Department of Cooperation and Cooperative Development, under the former Wheat Pool field man, the Hon. L. F. McIntosh. The staff of the new department was essentially that of the former Cooperation and Markets Branch of the Department of Agriculture with additions made necessary by the wider scope of activities and responsibilities. The duties and powers of the department were defined as follows:—

"The department shall take such measures as the Minister deems advisable or as may be required by the Lieutenant Governor in Council for the encouragement generally, of cooperative development in the province and in particular, but without limiting the generality of the foregoing, shall:—

(a) encourage and assist in the organization of cooperative enterprises among persons or groups who desire to provide themselves with, or to market, commodities or services or both on a non-profit, cooperative, self-help basis;

(b) provide for such inspection and examination of the affairs of cooperative bodies as may be necessary to secure the due observance of and compliance with the requirements of all Acts relating to cooperation and cooperative development and any regulations thereunder;

(c) institute inquiry into and collect, assort and systematize information and statistics relating to cooperation and cooperative development;

(d) establish a research service for inquiry generally into the operation of cooperative enterprise and for the making of such investigation and analysis of economic, social and other problems as may be deemed advisable for the encouragement of new or improved methods or means of cooperative organization and development and, in particular, to study and report upon questions relating to:—

   (1) agricultural production and the processing and marketing of agricultural products;
   (2) industrial development, manufacturing and sale by wholesale;
   (3) credit, investment and business finance generally;
   (4) retailing of goods and services;
   (5) community and other services;

(e) disseminate information relating to cooperation and

cooperative development in such manner and form as may be found best suited to encourage interest in the principles and practices of cooperation on a non-profit self-help basis;

(f) issue from time to time such reports, circulars and other publications relating to cooperation and cooperative development as may be deemed advisable;

(g) perform such other duties and provide such other services as may be designated by the Lieutenant Governor in Council." [9]

In carrying out its assigned duties, especially in matters concerning cooperative farms, the Department of Cooperation has worked together with the newly established Department of Reconstruction and Rehabilitation, and with the Adult Education Division of the Department of Education.

These activities have been kept up, in spite of a weakening of the C.C.F.'s hold on the government of the province. In the elections of 1949 the C.C.F. lost 14 seats, and in the ensuing cabinet reshuffle the Hon. T. C. Douglas, who continued as Prime Minister, assumed also the post of Minister of Cooperation.

Since the C.C.F. came to power in Saskatchewan, some of the legal impediments to cooperative farming have been or are in the process of being removed. Formerly an amendment to the Cooperative Associations Act of April, 1944, made incorporation of cooperative farms legally permissible. Now the rights of members in the wider community were clarified. Since the cooperative farm is paying taxes as one legal entity, only one vote was granted to all the members together in municipal and school elections. This restriction was corrected when the right to vote was made dependent on residence in a given area rather than on the paying of taxes. Still another

[9] First Annual Report of the Department of Cooperation and Cooperative Development of the Province of Saskatchewan, for the twelve months ended April 30, 1945, Regina, King's Printer, 1946, p. 9. f.

problem arose in connection with the Prairies Farm Assistance Act. By putting a cooperative farm on equal footing with an individual farmer, the Act deprives the members of their just claims. To secure each member in the cooperative rights enjoyed by the individual farmer, the matter was taken up with the Federal Government, and there are good prospects of remedy. Finally, as already mentioned, energetic attempts are being made to amend the Veterans' Land Act so as to make possible cooperative use of the grants.

The Department of Cooperation is taking seriously its task of research and education. Already in August, 1944, a consultation committee on cooperative farming, grown out of a conference of representatives of agricultural and veterans' organizations, was formed. A research committee was appointed which made a survey of cooperative farm practice in Canada and the U.S.A. and set down its findings and recommendations in a *Guide to Cooperative Farm Planning*.

Of a less spectacular, but of greater immediate value, are services offered to study-action groups. This work, too, is done in collaboration with the Adult Education Division. Pamphlets, some printed and some mimeographed, have been issued on the different aspects of cooperative farming during the last years. Space is generously granted to cooperative farm news in the Department's own bulletin *Cooperativ Development,* and in the Government's *Saskatchewan News.* Since the fall of 1946, a printed weekly study bulletin on cooperative farming is being sent out free of charge to individuals or study groups "interested in the possibilities of cooperative farming in Saskatchewan."

To keep in personal contact with all the groups, the Department has put a trained agricultural economist, Mr. Harold Chapman, in charge of an extension service for cooperative farms. Mr. Chapman visits the groups regularly and helps in an informal but efficient way to smooth the path of the new ventures.

A step forward in the consolidation of the development was marked by a meeting of delegates from the cooperative farms, held in Saskatoon, in October, 1946. This meeting, the first in a series to follow annually, was sponsored and attended by the senior officers of the Department of Cooperation, a representative of the Cooperative Union of Saskatchewan, and a delegate from the Dominion Economics Division. The meeting was to serve as an opportunity to exchange information, and to discuss the possibilities of forming a federation of all cooperative farms in the province. The delegates declared themselves definitely in favor of such federation which has since been established under the name of "Saskatchewan Federation of Production Cooperatives."

An Effective Demonstration

The number of cooperative farms in Saskatchewan is as yet small. Practically all of these farms are still in the stage of formation. But even in this initial stage they deserve attention and this for the following reasons:—

(1) They are not utopian. These farms are not being organized by religious or socio-reformistic zealots, who are bent on demonstrating the truth of a creed or of an ideology. They grow, rather, from the "grass-roots." These farmers resort to cooperation in the classical way, in response to essential needs that cannot be satisfied by individuals acting alone.

(2) These cooperative farms develop in a social and political environment which is sympathetic. They are as yet islands in the tide and ebb of competitive boom and bust. But they are not isolated, for there is continuity in them. They can be likened rather to peaks announcing in their rise the emergence of a new socio-economic continent.

(3) They mark the first coherent trend towards the coopera-

tive farm in a cultural pattern which was assumed to be adverse to anything but individual farming. Those who watched the U.S. Farm Security Administration's cooperative farms fail, and concluded that the American farmer was too much of a "rugged individualist" for such "experiments," may have to revise their conclusions. These Canadian farmers are in no way different from the U.S. farmer; they are only more hard-pressed. Their actions spring from the same motives that are present in the cultural background and the mental make-up of the American farmer. When the need for self-help arises, the pattern set by the farmers of Saskatchewan will be more acceptable to the American farmer than the kind of coopera-tive farm which has developed in Israel, Mexico, or in Soviet Russia.

(4) Even in their initial stage, these farms reveal the basic use to which cooperation can be put in rural economy. If nothing else, they demonstrate how problems of economic security, of a satisfactory social life, of succession on the farm, can be attacked successfully by cooperative group action.

(6)

THE CASE OF THE CHINESE INDUSTRIAL
COOPERATIVES

The industrial cooperative—known in England under the name of "productive society"—may be called the problem child of the cooperative movement. In the plans of the Rochdale weavers it figured as one of the main objectives. But when the English Cooperative Wholesale Society became strong enough to set up its own factories, it found that they would be more efficiently run by hired workers. The relatively few worker-owned shops or factories that sprang up have made a poor showing. They have always remained a subject for controversy which, by no means, can be considered as closed.[1]

Considered in the light of this controversy, the Chinese Industrial Cooperatives (C.I.C.) present themselves as an interesting piece of factual evidence. These cooperatives came into existence during the years of World War II, as part of the response to the destruction wrought by the invading Japanese. First organized in 1938, their number rose within the next three years from 69 to 1,867 societies and from a membership of 1,149 to 29,284. But in 1941 their growth was arrested, and with the end of the war their number declined quite considerably.

Taking China as a whole, the material impact of this development can hardly have been far-reaching. But the C.I.C. have, nevertheless, caught the imagination of cooperators everywhere, and even of some of the general public. The American Committee In Aid of Chinese Industrial Cooperatives (Indusco, Inc.), organized in 1938, succeeded in collect-

---

[1] J. B. Taylor, "Industrial Co-operation: Its Scope and Place in the Movement," *Review of International Co-operation* (January 1947).

ing during the nine years of its existence the sizeable sum of $3,600,000.[2] The reason for the strong appeal of the C.I.C. was mainly moral. It had certainly something to do with the typical American sympathy for the "underdog." In the case of China, this sympathy was not altogether free of misgiving. China, after all, was so much bigger than Japan that its humiliation could hardly be accounted for by a disadvantage in size. The suspicion could not altogether be suppressed that it was due, at least in part, to rot and corruption. It was therefore that the pluck and determination displayed by the Industrial Cooperatives were greeted with enthusiasm as symptoms of China's recuperation.

But, as far as cooperation is concerned, there was also a more objective reason for the interest aroused by the C.I.C. The industrial cooperative, much disputed elsewhere, appeared to have "made good" in war-torn China. What did this accomplishment signify? Was it temporary, or of an enduring character? Was the Chinese Industrial Cooperative merely a device born of the emergency of war, to be discarded at its end? Or did it respond to deeper needs, and was thus destined to become a lasting and significant component of China's economy? In the first case the experience of the C.I.C. would repeat, with slight variations, an already known pattern; in the second case, our opinion of the productive society would have to be revised.

But whatever the case, a sober answer to these questions can be found only in a factual review of the C.I.C. story.

China's economy is predominantly agricultural. "Almost 80 per cent of the country's 400 million people depend upon the land for their livelihood."[3] For the vast majority of the popula-

[2] Chen Han-Seng: *Gung Ho. The Story of the Chinese Cooperatives.* American Institute of Pacific Relations, Pamphlet No. 24 (New York, 1947), p. 46 f.

[3] Owen L. Dawson, "Agricultural Reconstruction in China," *Foreign Agriculture,* VII (June, 1943), p. 123.

tion the kind of livelihood derived from this predominant occupation is one of penurious subsistence even in normal times.

Since 1927, the National Government had tried to remedy this condition. The fostering of cooperatives was one of the measures taken. In China cooperation was used as a means of rural rehabilitation. Up to 1937 there were established 46,983 cooperative societies, counting 2,139,634 members.[4] It is interesting that the outbreak of the Sino-Japanese war strongly stimulated their growth. By 1938 their number had jumped to 64,565 and the membership to 3,112,629. During the following years the number rose steadily and reached, in 1941, a total of 107,904 societies and 5,979,212 members, in Free China alone. By July 1943, there were 165,018 cooperative societies of various kinds, with a combined membership of 11,871,809, and a capital of Chinese Dollars 187,378,934.[5] An attempt to coordinate the activities of the various cooperatives was made with the establishment of the Cooperative League of China, February, 1940.

Preceding the appearance of the industrial cooperative, a signficant shift took place in the development of Chinese cooperatives. Up to July, 1941, more than eighty-five per cent of the cooperatives were credit societies. (See Table I.) But by July, 1943, their percentage had sunk to fifty-three per cent, while agricultural production consumers' and marketing cooperatives gained ground. (See Table II.)

[4] Miachen S. Shaw, "Toward Planned Co-operative Economy," published by the China Information Committee in Chunking, quoted in "China—Recent Trends in the Development of the Co-operative Movement," *I. L. O. Cooperative Information*, No. 1, 1942.

[5] "China—Progress under the Three-Year Plan for the Expansion of the Co-operative Movement," *I. L. O. Cooperative Information*, No. 3-4, 1944.

TABLE I

DISTRIBUTION OF COOPERATIVES
BY TYPES UNTIL JUNE, 1941 [6]

Year	Credit %	Supply %	Production %	Marketing %	Consumption %	Misc. %
1939 (Dec.)	88.36	0.36	8.51	1.76	0.55	0.46
1940 (Dec.)	87.00	0.46	8.78	1.96	1.41	0.39
1941 (Dec.)	86.21	0.51	9.30	1.97	1.63	0.38

[6] "China—Recent Trends," *op. cit.*

TABLE II

DISTRIBUTION OF COOPERATIVES
BY TYPES IN JULY 1943 [7]

Credit 53.0%
Supply 6.5%
Agr. prod. 12.4%
Ind. prod. 4.9%
Consumer 9.0%
Marketing 10.0%
Utility 2.4%
Insurance 1.8%

[7] "China—Progress under the Three-Year Plan," *op. cit.*

As can be seen from Table II, the industrial cooperatives form about five per cent of all cooperatives.

Although one wool-weaving cooperative was established by members of the Nanking University, as early as 1935,[8] the real beginning of the C.I.C. must be set at 1937, the year in which the invading Japanese destroyed ninety per cent of China's large-scale industries.

Prior to the concentration of mass production in the great ports and river cities, China's industry had been exclusively one of small hand industries. "Towns and cities clattered with the activity of tiny shops. In each a few craftsmen using crude

[8] Ruth Weiss, "Chinese Industrial Cooperatives, Chengtu Report 1940," C.I.C. Hongkong Promotion Committee, 1941, p. 3.

tools, fashioned marvels of art or utility. Looms that looked as if they have come together by accident produced miraculous silks. Transport difficulties forced upon each area an independent economy." [9]

But when the "industrial revolution" reached China, these primitive industries were deprived of their economic basis. "In many localities the old hand industries languished, in some they died." Mass production put them out of existence. Thus, when the large-scale industries themselves were destroyed at the outbreak of the war, there was practically no other kind of industry to fall back upon.

In this situation cooperation, known and practiced in rural areas, appeared to offer a way out. A cooperative type of industrial production, decentralized, mobile, and dependent on small groups of workers, was called into existence, and was propagated under the slogan "Gung Ho"—Work Together.

The idea, unlike that of the consumers' cooperatives, did not come from the common men themselves. Credit for originating the C.I.C. is variously assigned to the American journalist Edgar Snow, his wife (Nym Wales), and the New Zealander Rewi Alley. All three were foreigners, and professionals. The one who undoubtedly was mainly responsible for organizing the movement was Rewi Alley who, as a factory inspector for the Municipal Council of Shanghai, had had ample opportunity to acquaint himself with the unbelievable misery of the Chinese industrial workers.[10] To him, the destruction of the Shanghai sweatshops appeared as no outright evil. It at least had the relieving aspect of offering an opportunity to replace sweatshop conditions with more decent forms of industrial enterprise. What Rewi Alley envisaged was a "small decentralized industry" spread over the towns and villages. And he

[9] "Co-ops Aid China," *Business Week*, February 8, 1941.
[10] Edgar Snow, "China's Blitzbuilder, Rewi Alley," *Saturday Evening Post*, February 8, 1941.

aimed at a rejuvenated industry in which "labor and capital could really work together," an industry "that could not only win the war but also win the peace." [11]

Among the immediate tasks which the C.I.C. set themselves, were a "speedy wartime reconstruction," a "productive refugee relief to supplant soup kitchens," "the training and spiritual mobilization of labor," and "quickly opening up the hinterland to defense against conquest." To these tasks had to be added that of educating labor for "duties of citizenship" and of "reconciliation of industry with the fresh air of the countryside."

Rewi Alley found ready support for his ideas in influential Chinese circles. In 1938 a committee was formed, consisting of Alley, the banker Hsu Sing-Loh, Frank Lem, an American educated engineer, Lu Kuang Mien, a cooperator trained in Scotland, Edgar Snow and his wife, and others.[12] Through the good services of Sir Archibald Clark Kerr, the then British Ambassador to China, the committee won the support of Generalissimo and Mme. Chiang Kai-Shek, and of Dr. H. H. Kung, Vice President of the Executive Yuan and Minister of Finance which granted the committee a loan of U.S. $500,000.

Alley divided all Free China into five areas, and the first cooperative, one of blacksmiths, was formed in Paochi in 1938. At the end of 1938 there were already sixty-nine industrial cooperatives in existence, with a membership of 1,149 a subscribed share capital of Chinese $16,292, and a paid-up capital of Chinese $10,206. By June, 1941, the number of cooperatives, organized in seven regions, had risen to 1,867, and had a membership of 29, 284. The subscribed share capital amounted at that time to Chinese $11,835,793 and the paid-up capital to Chinese $1,357,858. (See Tables III and IV.) From this time on, both the number of societies as well as of mem-

[11] *Ibid.*
[12] "A Nation Rebuilds—The story of the C.I.C.," Indusco, New York, 1943.

bers has steadily declined. The overoptimistic goal, set by the committee, of 30,000 societies with a membership of one million appears now further from achievement than ever.

TABLE III

GENERAL STATISTICS OF C.I.C.[13]
December 1938 to June 1942

Year	No. of Societies	No. of Members	Share Capital	
			Subscribed $	Paid Up $
1938, Dec. 69		1,149	16,292	10,206
1939, June 724		9,534	163,188	91,842
1939, Dec. 1,284		15,625	416,108	236,122
1940, June 1,612		21,330	714,996	488,214
1940, Dec. 1,739		25,682	1,219,347	843,245
1941, June 1,867		29,284	1,835,793	1,357,858
1941, Dec. 1,737*		23,088	2,348,084	1,972,204
1942, June 1,590*		22,680	5,645,558	4,553,392

Note: All figures in National Currency (Chinese), about $19.75 to one U. S. dollar, at that time.

* Consolidation of cooperatives into larger units, and losses due to inflation, brought figures down late in 1941 and 1942.

TABLE IV

C.I.C. STATISTICS, JUNE 30, 1942 [14]
Classified by Regions

Year	No. of Societies	No. of Members	Share Capital	
			Subscribed $	Paid Up $
Northwest	325	4,019	1,214,715	728,194
Chuan-Kong	247	4,800	2,194,775	1,921,432
Southeast	433	5,395	715,755	572,963
Southwest	246	3,485	408,868	327,055
Dien-Chien	158	2,497	839,324	785,124
Ching-Yu	118	1,610	183,748	167,217
Che-Wan	63	874	88,373	51,407
Total 1,590		22,680	5,645,558	4,553,392

Note: (1) The provinces included in these Regions are as follows:
Northwest—Shensi, Kansu, Nighsia and Chinghai
Chuan-Kong—Szechuan and Sikong

[13] *Ibid.,* p. 27.
[14] *Ibid.,* p. 28.

Southeast—Kiangsi, Fukien and Kwangtung
Southwest—Hunan and Kwangsi
Dien-Chien—Yunnan and Kweichow
Ching-Yu—Shansi, Honan and Hupeh
Che-Wan—Chekiang and Anhui

(2) The figures given above are based upon reports of the Regional Headquarters for June, 1942. These need to be amplified by data from other sources.

(3) Data in connection with the army blanket program is not included in the figures given above.

(4) All figures in National Currency (Chinese).

At least seven people, working in the same field of production, had to get together to start an accredited industrial cooperative. The registration of a cooperative was preceded by a study made by the Regional Headquarters. This study served to establish the need for the project, what equipment and raw materials it required and the costs involved. If the proponents were found to be serious the loan was granted, and work could begin. The amount, advanced from a revolving fund, varied from as little as U. S. 50.00 to as much as U. S. $7,000.00. Loans for shop equipment ran for five years, at a charge of six per cent interest. The interest on loans used for purchase of raw material was eight per cent. In many cases no other security could be obtained than the word of the members. The losses, nevertheless, were said to be few.[15]

Every member had to acquire at least one share in the association, valued at Chinese $2.00, of which five cents had to be paid on joining. No one, even if he owned a maximum of twenty per cent of the share capital allowed, could exercise more than one vote. The members elected from among themselves a chairman, who acted as their foreman, and determined their own hours of work as well as the rate of pay. Regular meetings were held in which current problems and the plans for production were discussed. Any violation of rules by a member was subject to penalty decided upon by all other mem-

[15] "Co-ops Aid China," *op. cit.*

bers. Expulsion for non-cooperative conduct was possible only by majority vote. Members who resigned received only the nominal value of their share holdings.

At the height of the development, the membership of the C.I.C. was composed mostly of refugees from areas devastated by war, wives of soldiers, "war widows," and disabled soldiers. There were "guerilla cooperatives," in close contact with the frontline fighters for whom they produced arms and other equipment, particularly blankets. All religious creeds were represented. There were Moslem cooperatives, cooperatives sponsored by Catholic priests, Protestant missionaries, and the Y.W.C.A. One cooperative was formed by patients of a Buddhist hospital who asked for funds to start a flour mill.[16]

Many of the war refugees went home after the collapse of the Japanese armies. This was one reason for the dissolution of many of the societies. But the membership of the remaining cooperatives probably remained of the same character as during the war.

The conditions of work were still most exacting. To escape bombing, the shops had often to be set up in dugouts, in caves in the loess hills, in peasant huts, or abandoned temples. No chimneys could be erected, and the motors used were mostly internal combustion engines driven by gas from smokeless charcoal. Machinery had to be small and easily removable. In many cases, the dismantled machine had to be carried across long stretches of territory, to be put together at a new point, temporarily secure from the enemy. Japanese soldiers had orders to shoot on sight any person suspected of being a member of the C.I.C.

To be rightly appreciated, the achievements of production listed in Table V must be viewed under the perspective of the circumstances described.

[16] "Industrial Cooperatives Among China's Refugees," based on material furnished by Dr. J. Henry Carpenter, *Information Service,* XXII, (September 25, 1943).

TABLE V

C.I.C. STATISTICS, June 30, 1942[17]

Classified by Industries

Industries	Number of Cooperatives									No. of Members	Loans Outstanding $	Monthly Production $
	West North	Kong Chuan-	East South	West South	Chien Dien-	Yu Ching-	Wan Che-	Total	%			
Machine and Metal Works	12	7	20	6	4	3	5	57	3.6	1,011	1,600,786	1,458,340
Mining	73	8	21	1	8			111	7.1	972	196,836	42,883
Textile	101	141	44	142	97	45	14	584	36.7	10,449	5,233,985	12,157,056
Tailoring	32	20	35	22	15	23	12	159	10.0	1,718	1,209,852	2,768,038
Chemical	40	46	160	31	13	22	10	322	20.2	4,494	4,083,906	3,310,663
Food Stuff	15	6	25	5	7	9	3	70	4.4	707	610,965	1,008,249
Stationery Supplies	7	6	17	1	4	6	2	43	2.7	749	929,090	901,431
Carpentry and Masonry	22	5	63	8	4	6	3	106	6.7	1,090	589,739	453,744
Transportation	2		2	3				7	0.4	67	46,750	15,400
Miscellaneous	21	8	46	27	14	1	14	131	8.2	1,423	1,225,948	1,907,140
TOTAL	325	247	433	246	158	118	63	1,590		22,680	15,727,857	24,022,944
%	20.4	15.6	27.2	15.5	9.9	7.5	4.0		100.0			

Note: All figures in National Currency (Chinese).

[17] "A Nation Rebuilds," op. cit., p. 29.

Table V indicates only the main fields of production; if all items produced by the C.I.C. were listed, the figure would rise to some 200. Among these the most important are clothing, shoes, blankets, soap, porcelain, leather goods, chemical products, medical gauze, machinery, power equipment, minerals, transport facilities and military material.

According to a note in the Indusco Bulletin of March 1944, Chinese Industrial Cooperative production was divided at that time as follows: "39% textile, 22% chemical (including paper-making), 10% clothing and shoes, and the balance hardware, foodstuffs, machines, cigarettes, and hundreds of consumer items."

To the purely economic achievements must be added accomplishments in research and in technical training. Among the more remarkable results produced by the five chemical laboratories must be listed methods developed to extract tannin from local materials, to make sugar, to manufacture paper from bamboo pulp, and to produce a substitute for caustic soda for the manufacture of laundry soap.

Technical training was often connected with the marketing of the products. For example, cooperatives that produced textile machinery offered free instruction in weaving to the buyer.

The individual reward was, at all times, quite scant. In some cases there was no cash profit, as in the coal mining cooperative (in the Tsing Ling Range) of which it was said that "The members work hard in return only for their food, but it is hoped that wages of Ch. $70.00 [U. S. $3.50 at that time] per month soon will be paid."

Wherever profits accrued, they were divided, at the end of the year, as follows: twenty per cent to the Reserve Fund, ten per cent for salaries (director and staff of the Federation), ten per cent to the local Industrial Cooperative Development Fund (for shares in the Federation), and ten per cent to the "Common Good Fund." The remaining fifty per cent was distributed

among the members who normally used twenty per cent of
their earnings for the purchase of shares in their own cooperative. The supervision of all financial matters lay in the hands
of a committee of directors elected from among the members
themselves.

TABLE VI

TABLE COMPARING PRIVATE AND
COOPERATIVE PRODUCTION

	Private	Cooperative
Annual production per loom (bolt)	250.1	258.4
Annual production per worker (bolt)	63.8	64.0
Net income per worker or member	10,210.00	20,030.00
Sales price per bolt	763	705
Material cost per bolt	508	464
Manufacturing cost per bolt	52	61
Net profit per bolt	38	43

Of particular interest in the above enumeration is the "Common Good Fund." This fund served to finance the social activities of the C.I.C., particularly Health Care and Education.

Contributions to the "Common Good Fund" were used to
establish health insurance, clinics, and hospitals. All medical
treatment was offered, so far as available, free of charge to
members of the C.I.C.; non-members had to pay a small fee.
The drugs and other hospital supplies used were partly produced by cooperatives. In March, 1944, altogether five hospitals and twenty-three clinics were run by the C.I.C.

In the educational field, the "Common Good Fund" helped
to establish reading rooms, to launch programs of adult education, and to set up kindergartens.

But the scattered efforts sponsored by the "Common Good
Fund" could not satisfy the need for more intensive education.
The hastily recruited members had only a too uncertain view
of the broader aspects of cooperation. As long as the loans kept

coming in, the cooperative rules were observed, more or less faithfully. But they were apt to be disregarded as soon as financial difficulties arose. It is true that violation of cooperative principles and "doubtful business practices" could be punished by dissolution. But each cooperative dissolved meant so much wasted money and effort.

Improved leadership offered a more positive solution to the problem.[18] Systematic training of leaders became the next urgent concern of the movement. It was organized in the so-called Bailie schools.[19]

The first full-sized Bailie School was established in Paochi, the birthplace of the C.I.C., in 1941; by 1943 nine schools had been established, located in the Southeast, Northwest, Chuan Kong and in the Northern Front region of the C.I.C. The largest of these schools was that at Shuangshihpu, with about eighty students; the others averaged about forty students. All these schools were founded with the assistance of the "Institute for the Advancement of Chinese Industrial Cooperatives." In 1947, only one, the Shantan or Sandan School, was still functioning, and this under great financial difficulties. Sandan had close to two hundred students, of whom eight were girls.[20]

The program of Sandan provided for a basic course extending over two years. During the first year, the curriculum was partly practical, partly theoretical. The emphasis during the second year was on practice, in the workshop of the school, or in one of the cooperatives.

[18] George A. Hogg, "Training Cooperative Leaders for China," reprint from *Free World,* June, 1943. (George Hogg died from a tetanus infection which he caught while helping to set up the Shantan school.)

[19] These schools were named after Joseph Bailie, a missionary who advocated industrial education for the Chinese. He won, in the late twenties, the support of Henry Ford, who agreed to offer training to a number of Chinese youth at his factories in Detroit. Some of these young men helped later Rewi Alley in his engineering work.

[20] Chen, *op. cit.,* p. 53

The students ranged from twelve to eighteen. They were carefully selected, and only those who offered every promise of becoming leaders of the movement were admitted. The students tried to apply the principles of cooperation at the training stage. They organized their own consumers' and credit cooperatives. The leaders of the C.I.C. based a great deal of their hopes upon the good experiences had, so far, with the personnel produced by the Bailie schools.

As we have seen, the C.I.C. were inaugurated from above. This fact could not but influence to a large degree their character. In the first stage of the development all control rested with the "Headquarters" and the "Depots." A genuine cooperative spirit could hardly develop. A change in the setup indicated itself, therefore, as soon as the leaders realized that the membership had acquired the degree of cooperative maturity which made such rigid control from above superfluous and damaging.

In June, 1943 a tnorough reshaping of the organizational structure took place. Subsequently, the cooperatives became completely autonomous in the administration of their own affairs. A number of cooperatives in one area, usually between ten to thirty, formed so-called Federations. Each Federation was a financially independent unit, with a common treasury for all affiliated cooperatives. These Federations handled the marketing of goods and the purchase of raw materials for their affiliates. The officers of each Federation were elected by the cooperative membership, and served during a specified term.

The Federations directed also all cultural activities of their members, and served as connecting agencies between cooperatives of different areas. According to "A Nation Rebuilds," there were in 1943 altogether thirty-three Federations in all Free China. Eight in the Northwest region, nine in the Szechwan-Sikang region, five each in the Kiangshi-Fukien-Kwantung and Honan-Kwangsi regions, and three each in the Shensi-

Hoan, and Yunan-Kweichow regions. By 1945 their number had been reduced to twenty-six; ten Federations in the Northwest comprising 174 affiliated cooperatives; eleven in the Southeast, with 118, and five in the Southwest with 78 affiliated cooperatives.[21]

The Central C.I.C. Headquarters and the seven Regional Headquarters were, since 1943, divested of their original directive function, and played from then on a purely promotional role. A further promotional agency was the International Committee, which solicited funds from abroad. Affiliated with this Committee were the already mentioned Institute for the Advancement of Chinese Industrial Cooperatives, and Indusco Services, an agency operating in the field.

The function of the Institute for the Advancement of Chinese Industrial Cooperatives was mainly that of stimulating and supervising industrial research and experimentation. The Institute subsidized the Bailie Schools; it also organized regional training classes; it carried on economic research for the benefit of the cooperatives; and it published pamphlets and periodicals.

Indusco Services, finally, headed by Rewi Alley, was the agency directly concerned with the organization of new cooperatives. It furnished the technical aid and advice needed in the initial stage of development; it helped reorganized cooperatives whose production lagged, and it investigated the productive potentialities of each area in which cooperatives were planned.

Are we dealing here with a phenomenon that would not and could not outlast the emergency which produced it? Or does what we have learned about the C.I.C. warrant the conclusion that they are destined to play a lasting part in any Chinese economy?

It is true that, as we have seen, the rapid growth of the

[21] *Ibid.,* p. 48.

C.I.C. came to a stop already in 1941. From then until the end of the war the decline in numbers was small and partly due to consolidation. But how did the end of the war affect the C.I.C.? Their number fell from 1,590 societies with 22,680 members in 1942 to 282 societies with 3,518 members. And even though these figures, quoted by *Cooperative Information* (No. 8, 1947), referred to the Kuomintang area alone, they mean only one thing: a major crisis. The reasons given by the same source for this crisis were: insufficient equipment, inflation, lack of capital, competition from the import of foreign goods, and loss of membership through repatriation of refugees. To this list we can add from another source[22]: unsettled conditions, loss of wartime markets, and adverse political conditions.

Shall we decide, then, that the C.I.C. was merely a wartime creation, and dismiss it as such? In the then Communist controlled part, the picture appeared to be quite different. There, according to Chen Han-Seng, the C.I.C., after a slow start, began to spread at exactly the same time, 1942, that their growth was arrested on the other side. Their number kept growing even under Civil War conditions and reached a figure of 590 societies by September 1946.[23]

Thus it would appear that "political conditions" rather than Japanese invasion were the decisive factor in the fortune of the C.I.C. So long as these conditions remain unsettled, no definite answer can be given to our first question.

But does the fact that we do not feel entitled to dismiss the C.I.C. simply as a wartime expedient imply that we have to revise upward our opinion of the industrial cooperative and, by implication, of the productive society?

In order to answer this question, we have to take a close look at the real nature of the C.I.C. One thing that we dis-

[22] *Ibid.*, p. 55.
[23] *Ibid.*, p. 58.

cover upon doing so is that the C.I.C. was essentially not at all the same thing as the productive society. Both, it is true, adhere to the basic cooperative principles and have in common such features as self-government and copartnership of the workers, and the sharing of responsibility and control by the members. But they differ in one fundamental aspect. The initial capital of a productive society is put up, as a rule, by the workers themselves. The C.I.C., on the contrary, were neither initiated by the workers themselves, nor were the founding members in a situation to make more than a token contribution to the capital investment. The capital of the C.I.C. came, as we have seen, partly from Government, and partly from private fund raising agencies. Even the field of production, e.g. army blankets, was not chosen by the workers but was assigned to them from above. All they themselves put into the pool was their capacity for labor. They worked cooperatively and they shared the profits equitably; but their dealings were essentially those of a group contracting for labor.

Seen in this light, the C.I.C. were not real industrial cooperatives or productive societies, but rather "Collective Labor Contract" Associations, of one kind with the old Russian "Artel," or the Italian masons or bricklayers collectives, the "Bracchianti" and "Muratori." [24] Thus their case is one of an interesting, though not altogether novel, application of limited cooperative techniques to a given emergency situation. Their experience, however, has hardly any bearing on the issue of the industrial cooperative itself.

[24] Sidney and Beatrice Webb: *The Consumers' Co-operative Movement,* (London, 1921), p. 477.

(7)

PAINS OF GROWTH IN ISRAEL

The Kvutza Faces Change

Of the three main instances of modern cooperative community —the *kolkhoz*, the collective *ejido*, and the *kvutza*—it is the last which has succeeded, in the most unqualified manner, in realizing the original objective of those who gave the cooperative movement its propitious start. The *kvutza* has accomplished what the Rochdales Pioneers themselves defined as their ultimate goal: "That as soon as practicable, this society shall proceed to arrange the powers of productive distribution, education, and government, or in other words to establish a self-supporting home colony of united interests, to assist other societies to establish such colonies."

Not that this accomplishment has made the *kvutza* immune against all perplexity. Since it first came into existence, some forty years ago, the *kvutza,* and later the *kibbutz,* has grown into a substantial development, comprising today over 200 settlements with more than 60,000 souls. Although still growing, the *kvutza* and the *kibbutz,* because of certain inner and outward changes, find themselves at present facing a number of problems which, to some, appear to have the impact of a crisis. What is the nature of these problems?

A most complete, if somewhat discursive, inventory of puzzling issues can be found in several articles published in *Mibifnim*[1], particularly Eliezer Garbitzki's "Bemivhan Hatekufa" (On Examining the Present Situation). According to these papers, the issues seem to fall quite naturally into two main groups: those caused mainly by internal changes, and those arising chiefly from the settlements' relation to the new

[1] *Mibifnim.* Ain Harod. Published by Histadrut, in Hebrew. III, 4 (April 1949).

political situation. The problems of an internal kind appear, in turn, to have two main sources: (1) social and economic solidification, and (2) growth in size. The external problems appear to be related to: (1) the establishment of the State of Israel, and (2) the requirements of unrestricted immigration.

Problems of an Internal Character

Under the heading of problems arising from the social and economic solidification of the *kvutza* and the *kibbutz* comes, first, the increasing demand for individual comfort. So long as the settlements were in the formative stage, struggling for mere survival, the members hardly thought of making any such demands. Everyone knew that he was in it not for personal gain, but in the interest of a cause that could not be served without sacrifice. With the economic and social gains of the settlements came the feeling that the individual, notwithstanding all devotion, now had a right to claim part of the bounty for himself. Members who find that the group does or cannot satisfy such claims, sometimes will resort to practices which violate the spirit of genuine equality. Some, for instance, when sent on outside work, save on expenses and buy special toys for their own children. Or, some who have no patience to wait until the *kvutza* is able to afford a radio for each room, have radios sent to them by outside friends or relatives.

Another set of problems arises from the growth in size of the *kvutza* or *kibbutz*. Even those settlements which were in principle opposed to large size have today grown far beyond the intended scope; others are reaching the proportions of small townships. The branches of economic activity have expanded and become highly specialized; they can be managed only by skilled and experienced foremen. In spite of such growth, the original methods of administration, satisfactory for a small group, have remained essentially unchanged. In consequence, the General Assembly has lost more and more of its important democratic function. To take a stand on any issue requires specialized knowledge, which only a few of the members possess. The majority has to leave decisions to the expert few who, in the interest of the community's economy, must be trusted with the conduct of affairs. Elections, instead of being truly democratic and free, turn into more or less automatic confirmations of those technically qualified for office. A "managerial class" has developed, and the way is open for personalities stronger in will and intellect than in ability for cooperation to gain ascendence. Trends of this kind, left unchecked, tend to enhance apathy on the part of the membership, to widen the gap between leadership and the rank and file, and to undermine the democratic foundations.

Another important problem arises in connection with what may be called "careerism." As has been pointed out elsewhere[2], there exists in the *kvutza,* in spite of complete economic equality, the equivalent of what in our society we call "making a career for oneself." In the *kvutza,* such a career is, to be sure, not related to one's bank account or any material possessions, but to a scale of values based on work-performance. That there is such a thing as differential status among the members of the *kvutza* or *kibbutz,* a simple sociometric test will easily reveal. Again, so long as everyone was part of a desperate struggle for survival, little attention was paid to such differences in social status. Today, with increasing solidification, most of the members find themselves to be middle-aged, and inclined to take stock of themselves. Those who have achieved some prominence, whether in the *kibbutz* or outside or both, tend to feel pleased with themselves. Others, however, who have real or imagined ground to think themselves outranked, tend to develop a sense of frustration. This appears to be particularly grave in the case of women who, generally,

[2] H. F. Infield, *Cooperative Living in Palestine,* ch. III.

in the formative stage, represented the most active and integrating element in the *kvutza*.

In the light of the often tenacious resistance to schemes of
cooperative living offered by women elsewhere, the role of the
kvutza woman was unique in this respect. Today, according to
some observers, it is the woman who, on finding her ambition
for a career equal to that of man thwarted, becomes the cause
of many petty irritations within the *kvutza*. When couples
leave a settlement together, it is more often than not the
woman who has prompted the separation. Needless to say that,
if this observation is true, we witness here the making of a
problem fraught with the gravest consequences for the *kvutza*.

Problems of an External Character

Equally troubling are some of the problems directly connected with the establishment of the State of Israel. The communal settlements provided the *Haganah,* and more recently
the army, with its most substantial material and human contin-
gents. This has put a great strain on the settlements' resources
and deprived them of many of their most active and experienced members. The drain is still making itself felt sharply in
the economic and social life of many a community. The settlements are coping valiantly with the exigency, and are willingly
shouldering their part of the burden. What puzzles them, however, even to the point of resentment, is an apparent tendency
on the part of the government to snub the settlements when it
comes to vital political decisions. The settlements feel that
without them there would have been no State of Israel. They
expected the government to make widest use of the settlements'
experience in self-government, if not to base itself altogether
on the *kvutza's* tested principles of operational, or shall we say
"co-operational", democracy. What appears to them to be
happening instead is that the government of the new state, like
governments elsewhere, has adopted political expediency as its

guiding principle, even attempting to interfere with the settlements' internal autonomy. The defense of this their most
cherished prerogative against incursions, has become one of
the most galling problems for the *kvutzot* and *kibbutzim.*

A final source of irritation has been the unrestricted immigration policy pursued by the new state. To help settle new
immigrants had always been one of the most important and
most superbly accomplished functions of the *kvutza.* Formerly,
immigration was usually well prepared, both physically and
ideologically. It represented an element from which the *kvutza*
could derive additional strength. The new immigration lacks
such preparation. It consists to a large degree of people whose
only *hakhshara* (preparation) has been the suffering experienced in the Nazi concentration camps, or in the detention
camps for displaced persons. These people come to Israel
obsessed by hate of any kind of regimentation. To many of
them the communal settlements look just like another camp.
They are determined not to let themselves be pushed again into
"anything of this sort." The settlements, possibly for the first
time in their history, find themselves worrying about human
reserves, and this at a time of unprecedented increase in the
size of immigration.

Cooperative Movement and Cooperative Community

The problems besetting the communal settlements in Israel
today are undoubtedly grave. No wonder then that to some
they appear as symptoms of a fateful crisis. They lose, however, much of their catastrophic appearance if viewed in a
more detached manner.

A more objective perspective is gained if the *kvutza* is seen
as what, at least to this observer, it essentially is, a modern
cooperative community, and, as such, a component part of the
cooperative movement. The present problems of the *kvutza*

must be seen assessed in the light of the history of this movement. What has been, briefly, the role of the idea of the
cooperative community in the cooperative movement?

Sociologically, the cooperative movement may be characterized, in Herbert Blumer's terminology, as a "specific social
movement" [3]. Like other social movements, it started from a
"condition of unrest", when working men of England, in the
wake of conditions created by the Industrial Revolution, found
themselves incapable of satisfying even their most immediate
needs. Faced, on the one hand, by incredibly low wages, and
by high prices on the other, the worker and his family were
forced to vegetate on a substandard level. Instead of resigning
themselves to their fate, they began to grope for a way out of
an unbearable situation.

In the first, the "groping" stage of the movement, the working men found two leaders, neither of them belonging to the
working class themselves. One was Robert Owen, a wealthy
textile manufacture from New Lanark; the other Dr. William
King, a physician practising in the city of Brighton. Of the
two, Robert Owen was undoubtedly the more dominant personality. His attitude towards the workers was paternalistic,
and his ultimate goals were somewhat vague. He aimed at a
total reformation of society, a complete "new order of things."
As the main device to bring about that "new order", he proposed "villages of cooperation." Among his attempts to establish such villages, the most famous was "New Harmony,"
founded in Indiana, U.S.A., in 1825. The enterprise ended in
failure three years later. Other experiments in the U.S.A. and
England met with a similar fate, and Robert Owen began to
concern himself with more immediate possibilities of self-help.
At this point Dr. King's influence became prominent in the

[3] Herbert Blumer, *Handbook of Sociology* (New York: Dryden Press,
1941), p. 258 ff.

movement. Against Owen's "villages of cooperation," Dr. King favored the cooperative store. He focused his and his followers' attention on the most efficient way of organizing such "union shops." By 1830, their number had risen to some 300. But four years later, in 1834, virtually all of them had gone out of existence.

The Second Stage of the Cooperative Movement

The second phase of the cooperative movement was ushered in by the famous Rochdale Weavers. That phase was characterized by what may be called "formalization." The story of the movement's success starts with the establishment of a consumers' cooperative store in the English town of Rochdale. The twenty-eight weavers who organized the store laid their plans very thoroughly. They defined, as we have seen, their long range aims in terms of a "self-supporting home colony of united interests," and they conceived of the consumers' store as the most effective means towards the attainment of these aims. Profiting from the mistakes of their predecessors, they evolved a set of ingenious and exemplary standards for the organization of their store. The "Rochdale Principles" have since become widely known and practiced. They were prolific enough to call forth a worldwide movement which, at the beginning of World War II, counted a membership of some 145 millions, in more than fifty different countries.

The consumers' store spread rapidly from Rochdale to other parts of the United Kingdom, to the European continent, America, and virtually every civilized country of the world. In the process, the movement grew both "horizontally" and "vertically." Learning to handle practically every known commodity, it grew also in size of membership, adding, in the so-called Wholesale, a purchasing and producing structure of its own. The relations within the movement became "institutionalized" and produced such unifying agencies as the different Cooperative Unions and Cooperative Leagues, and, on a wider scope, the International Cooperative Alliance. In Britain, even a Cooperative Party was formed.

The acknowledged dominant personality of this stage of growth was John T. W. Mitchell. From 1874 to 1895, as chairman of the English Cooperative Wholesale Society, he "taught the primacy of the consumer." [4] Carr-Saunders lauds him as "perhaps the only prophet produced by the movement, and certainly the greatest," because "He it was who gave meaning to the growing cooperative enterprise, who saw clearly that the profits of such enterprise, at whatever stage they arose, must return to the ultimate consumer if the ideal of direct production for use, in opposition to the system of production for profit, was ever to be achieved." [5]

The rapid growth of the cooperative movement, the spread of the consumers' store and other related societies, such as credit unions, marketing, purchasing, and processing cooperatives, seemed to justify the almost total neglect of the Rochdale Pioneers' original objective. A certain uneasiness, however, began to make itself felt among the more sensitive of the cooperative leaders. This uneasiness grew out of the increasingly apparent disparity between the material achievements of the movement and its negligible social effect. The sense of this disparity found its most eloquent expression in the writings of the Irish poet and cooperative leader, George Russell. In his *The National Being,* we find the following statement: "If people unite as consumers to buy together, they only come into contact on this one point; there is no general identity of

[4] Horace M. Kallen, *Decline and Rise of the Consumer* (New York: D. Appleton-Century Co., 1936).

[5] Carr-Saunders, Florence, and Peers, *Consumers Cooperation in Great Britain.* Third revised edition. (London: Allen & Unwin, 1942), p. 38.

interests. If cooperative societies are specialized for this purpose or that—as in Great Britain or on the Continent—to a
large extent the limitation of objects prevents a true social
organism from being formed . . . The specialized society only
develops economic efficiency." [6] Against this George Russell
proposed, as a remedy, a community in which "all the rural
industries are organized on cooperative lines."

Dissatisfaction with mere economic accomplishment has
prompted since some incisive criticism of the movement. Most
articulate is that made by Carr-Saunders and his associates.
After a penetrating analysis, they come to the conclusion that
the movement faces "a danger of stagnation" manifesting itself
in the following symptoms: (1) "a blurring of the features
which distinguish it [the cooperative business] from competitive private enterprise," mainly by adopting the "same criteria
of success;" and (2) "decline of sensitiveness on the part of the
leaders to the needs of the general body of consumers." Looking for the causes of these symptoms, they find them to be of a
twofold nature, partly external, and partly internal. Among
those of an external nature figures chiefly that of "State inter-
vention on the side of large producer interests." More ominous,
however, is the main internal cause of stagnation which is
described as the weakening of the "intense community feeling
for which the earlier forms provided an effective channel." [7]

The Present Stage

It appears that Russell and Carr-Saunders agree that, in
spite of the spectacular economic achievement, something
essential is missing in that second stage of the movement.
While using different words, they point—one explicitly, the
other by implication—to the same thing: to the idea, if not to

[6] George Russell, *The National Being* (New York: Macmillan, 1930)
p. 40.
[7] Carr-Saunders *et al., op. cit.,* p. 516 ff.

the practice, of the original objective of the movement, the "self-supporting home colony of united interests," or as we would say today, the cooperative community. It is interesting to note that Russell and Carr-Saunders were apparently unaware of the fact that the thing they were hoping for in the future had already come into existence. We refer to the establishment in Palestine, in connection with the Zionist colonization enterprise, around 1910, of the first modern cooperative community, the *kvutza* Dagania A.[8] We call that cooperative community "modern" because, unlike the "utopian," religious or socio-reformistic community, it grew, like the cooperative movement itself, out of some very tangible basic needs which, under the given conditions, could not have been satisfied adequately in another form. In other words, the *kvutza* has the same natural history as the other cooperative associations based on the Rochdale Principles, and is thus generically identical with them. The men who founded the first *kvutza*, had probably little inkling of the universal implications of their venture. Historically, their achievement marks a turning point: it marks the end of the second stage of the cooperative movement and the beginning of its third, the present stage.

This third stage is still in its inception. There are, however, abundant signs which justify the assumption that it will be dominated by the emergence of comprehensive cooperation, or the cooperative community, as a legitimate issue of the cooperative movement. The extent to which cooperative farming is employed as an instrument of rural reform in the Soviet Union and Mexico, and as an effective device for resettlement in Israel, is sufficiently indicative. Trends pointing in the same

[8] That neither of the two eminent students of cooperation was cognizant of this fact, is somewhat puzzling. It appears more understandable in the case of Russell whose book first appeared in 1916. Carr-Saunders' study, however, was published in 1938, when the *kvutza* had grown to considerable numbers and attracted world-wide attention of social scientists.

direction can also be found in Canada, India, Australia as well as Eastern Europe and elsewhere.[9]

Indicated Solutions

Against the background of even this sketchy historic elucidation, the problems of the *kvutza* lose much of their uniqueness. They are, rather, characteristic of the kind of problems generally encountered in the second stage of the cooperative movement. The root problems are generally related to internal changes, on the one hand, and to difficulties in the relation to the given political situation, on the other. Thus it would seem that the problems facing the *kvutza* today are quite natural to the movement of which it is part, and that they do not arise from the specific shortcomings of the *kvutza* itself.

Reassuring as this general observation might be, it does not automatically solve the *kvutza's* actual difficulties. It merely indicates the direction in which possible solutions might be found. Recourse to the original sense of community was, as we have seen, the remedy suggested by competent counsel against the "danger of stagnation" arising from segmental cooperation. The dangers, in the instance of the *kvutza,* facing comprehensive cooperation call, in principle, for the same kind of response, that is, a conscious revitalization of the creative impulses that were present at the *kvutza's* origin.

As to the specific problems besetting the *kvutza,* the following possibilities of solution suggest themselves:

a) *Regarding demand for individual comfort:* Some suggest greater attention on the part of the group as a whole to such demands of the individual. Others hold that demands of this kind, once accepted as legitimate, have a tendency to multiply infinitely, and may lead to sharing out, to the joint stock company. Therefore, they recommend a return to the simplicity

[9] Cf. the essay on "Some Recent Developments in Cooperative Farming" in this volume.

and modesty of the beginnings. Advocates of the latter plan might find actual support for their proposal in the experience of the Hutterite colonies which demonstrates that frugality has been a significant factor in the longevity of cooperative communities.[10]

The real issue is probably not one of either or—either no comfort at all or all the comfort one can afford. It is, rather, how much of individual comfort may yet be compatible with cooperative living. Thus far, it has been determined merely by way of trial and error. It would seem that more dependable, scientific methods of gauging the degree of compatibility would go far in helping to solve the problem.

b) *Regarding democratic participation:* The most effective way to meet the difficulties arising from the growth in size of the settlements would seem to be "decentralization." Dissection of the membership into units easier to manipulate, suggested by some, would hardly do. What seems to be needed is, rather, reorganization in such manner as to make the size of the settlement more commensurate with the cooperative values it aims at. Whether this can be accomplished without reduction of the large-size community into smaller settlements—which, again, would mean a return to the original *kvutza*—remains to be seen.

Under the mandate, land was scarce, and sub-colonization was hardly feasible. Today, with more land available than settlers, it should at least be worthy of consideration. If we turn to the Hutterites again, we find that they never let their colonies grow to more than 150 or, at the most, 200 souls;

---

[10] The first Hutterite colony was established in 1526. Today, some 9,000 Hutterites live in more than ninety such colonies in the U.S.A. and Canada. Based on religious principles, the Hutterite colonies are organized along lines virtually identical with those of the *kvutza*. For more details, see this writer's *Cooperative Communities at Work,* chap. II; "Sociology and the Modern Cooperative Community," and "The Link with the Past" in this volume.

and they have succeeded in surviving now for more than 400 years.

c) *Regarding "career" and the status of woman:* Problems of this kind, connected as they are with the natural factors of inequality of age and sex, cannot satisfactorily be dealt with on a purely rational level. Any attempt at solution will require a particularly great effort of clarification and ingenuity in implementation. Some lessons may be learned from the study of past and present arrangements in other communities. Here again, exact fact finding methods should prove of great help. One of the greatest obstacles to a solution, however, is inherent in some of the central concepts of *kibbutz* ideology, particularly, the cluster of tenets based on the postulate of unqualified equality of man and woman. It is only to be hoped that the will for survival will prove stronger than some cherished but unrealistic notions, especially as such notions are increasingly rendered obsolete by the settlements' own experience.

With regard to problems arising from the settlements' relations to the political situation, it must be remembered that the State of Israel has not been in existence long enough to permit any fair judgment about its long-range intentions towards the settlements. The suspicion that the new government intends to interfere with their internal autonomy seems to have no basis in fact. It is rather the ambivalent attitude of the *kibbutzim* themselves which threatens to undermine their autonomy. When the *kibbutzim* begin to hire paid labor for their factories, they cannot blame the government for suggesting that this labor be taken from along the new immigrants settled in the vicinity of the *kibbutzim*.[11]

[11] Such a proposal was actually made by the then Premier Ben-Gurion. It was vehemently rejected by the *kibbutzim* and subsequently abandoned. According to latest reports, however, the hiring of labor is becoming an accepted practice in many *kibbutzim*.

Regarding the difficulties arising from unrestricted immigration, only time will show whether it will be possible to do anything about them. Education of the new immigrants, if efficiently organized, may produce some useful results. Most of the new immigrants, particularly those of a more advanced age, are probably lost to the *kibbutz* movement. The question arises whether sources of a more desirable kind of immigration can be found. Understandably enough, the eyes turn to the largest remaining reservoir of Jewish population in North America.

Here there may in fact be found new and rewarding yields. However, the whole approach to American-Jewish youth will need considerable reformulation, both in concept and in practice. It is improbable that the foreseeable future will see the unleashing in America of such violent "pushes" as the Czarist pogroms, the Polish massacres, and the Nazi terror, which caused the several waves of mass migration to Palestine. Since the needs of American youths vary significantly from those of their European contemporaries, the appeal to them must be in different terms as well. The methods of preparation, particularly the *Hekhalutz* Training Farms in the U.S.A. will have to attain a level of efficiency far above that prevailing at present, if the results are to stand the "reality test" of Israel.[12] Improved methods of preparation are all the more important as the influence of American youth immigration will at best be qualitative rather than quantitative.

Vitality and Self-Examination

What the *kvutzot* and the *kibbutzim* need above all to overcome their present difficulties, is vitality and self-examination.

Nobody who has had recent opportunity to observe the communal settlements can doubt their vitality. In spite of all the

[12] Cf. "Testing of a Pioneer Training Farm" in this volume.

problems facing them, their number is still growing. They are busy and successful in improving their economic and social conditions. Most important of all, they do not have to worry about succession. Even though there may be some disquieting turnover among the older members, there is virtually no desertion among the second generation. The youths growing up in the settlements apparently see no reason for abandoning the ways of their fathers.

As to self-examination, the very fact of the critical articles in *Mibifnim* referred to above, all written by members of the settlements, testifies to the astuteness with which they tend to re-examine their own situation. That this examination is often naive and lacking in analytical skill or technical experience, is a formal shortcoming, which can be mended by professional advice and assistance.[13] Seriousness of will and purpose is certainly present.

As for the external problems, they are mostly beyond the settlements' control. Once its own position has become well established, the new government can be expected to recognize the settlements as one of the most precious assets of the State of Israel. It would then be eager not only to preserve them, but to do all in its power to provide the most favorable conditions for further growth. It might find that among the measures to serve this end, a policy of selective immigration would be very helpful.

There appears to be no reason to doubt the ability of the communal settlements in Israel to overcome their present day difficulties. The present "crisis" has all the hallmarks of a process which will result in some healthy clarification, particularly of the beclouded issue between exigencies of the

[13] It is to the social sciences that the settlements are entitled to look for such help. The recent establishment at the Hebrew University in Jerusalem of a chair for the cooperative movement is a welcome step designed to serve this purpose.

national economy and the postulates of cooperative living. Such clarification cannot but help fortify the new State of Israel, and add new strength to the cooperative settlements themselves.

(8)

EFFECTIVE EDUCATION: THE STUDY GROUP AND THE TRAINING FARM

Education and Cooperation

In a world still largely dominated by the ideology of competitive, so-called "free enterprise," resort to cooperative action represents a departure from accepted norms of behavior. For the individual, such departure will always imply a change of attitudes. The change will usually be initiated by dissatisfaction with the way things are being done, and by consequent desire for improvement; it will be followed by a search for a way of doing the same things better, and by finding such a way in cooperation; and the change will, finally, be accomplished by adopting the cooperative way of doing things, and by conforming to norms of behavior essentially determined by the Rochdale Principles. Whatever the extent of the field mapped out for cooperation may be, the process will in all cases require a certain amount of learning. Such learning will range all the way from the unlearning of old and often deeply ingrained attitudes, to the acquisition and activation of new modes of behavior. It is only when this process has been carried through effectively that cooperation can become more than a matter of mere economic convenience. Hence the stress laid by the Rochdale Pioneers on education.

In deference to the authority of these pioneers, every genuinely cooperative association has since made education an item on its program, and even of its budget. It is true that in most cases deplorably little that could be called effective has been done in practice. In spite of that, the cooperative movement has grown, and is still growing, by leaps and

158

bounds. This is probably due to the fact that the vast majority of cooperative associations still remains of the segmental type. Membership in a consumers' store, a credit union, a marketing or purchasing cooperative, affects the individual only in one of his many social roles, that of a buyer of groceries, borrower of money, trader of agricultural products or implements. The change required in this case extends merely to one segment of the individual's social behavior, and can be carried through with a minimum of effort. It also remains, it ought to be evident, extremely limited in its effect on the behavior of the individual as a whole.

Where, however, the type of cooperation practised requires a fuller participation of the individual, neither numerical nor, for that matter, any other marked results can be expected from such a minimum of effort. This is particularly the case in cooperatives which include production as well as consumption and which, consequently, tend to mould all other social functions of the group in the cooperative sense. Where multiple cooperation is the goal, or where the tendency is towards cooperative farming, to say nothing of the cooperative community, the required deep-going change will necessarily have to include more than one segment of the personality concerned. It will have to extend to most, if not all, of the individual's basic attitudes. It is obvious that a change of this kind can be produced only with a great deal of effort and education.

To-day, with quite a few governments introducing or contemplating the introduction of cooperative farming as the solution to their rural problems, the issue of education for agricultural cooperation of this kind assumes immediate urgency. Experience has shown that cooperative farming cannot be introduced by decree alone. Administrative measures prove not only often ineffective but, if rigidly enforced, may produce stubborn resistance, with all the dissatisfaction, conflict, and spiteful waste implied. Little else can be done to

avoid such undesirable reaction, it would seem, than to make
a sincere effort to enlist the voluntary participation of those
who are to benefit from the projected reform. To make this
effort succeed ways and means must be found to prepare the
minds of those concerned for the intended action. A means of
this kind, which has proved itself superior to any other form of
education for cooperative action, is the study group. It may be
useful, therefore, to consider briefly some instances of agricultural cooperation in which the study group has played a significant and apparently very effective part. We have in mind
particularly these three: (1) The socalled *Antigonish Movement* (2) the cooperative farm development in Saskatchewan,
Canada; (3) the cooperative community which developed in
Israel, the *Kvutza*. Since we intend to derive lessons generally
applicable to the *introduction* of cooperative farming, we shall
limit ourselves to some of the main educational aspects in each
case and to the study group in particular.

The Antigonish Movement

The education for self-help among the farmers, fishermen,
and industrial workers of the Maritime Provinces of Canada—
Nova Scotia, Prince Edward Island and New Brunswick—
originated from the Extension Department of the Catholic St.
Francis Xavier University. The fact that this University is
located at Antigonish in N.S., has given the development its
name. There are several men who can claim credit for having
initiated the *Antigonish Movement*. Chief among them are:
Dr. J. J. Tompkins, Dr. Hugh MacPherson, Dr. D. J. Mac-
Donald, Rev. Michael Gillis and a number of other men connected with the Extension Department. There have also been
some outside influences which have helped shape the course of
the development. Among these, the most important have been
the example of the Scandinavian countries, particularly the

Danish high schools, and a book, E. L. Thorndike's *Adult Learning*. But it was one man especially who gave the movement its impetus: Dr. M. M. Coady. It was he who, in 1927, as the newly-appointed head of the Extension Department—a job he held until recently—assembled 12 farmers in a rural section of Cape Breton in order to consider the possibilities of their own community. Out of the serious discussions that were held these farmers conceived ideas of new lines of production, and proceeded to carry them out, with surprising success. The characteristics of the main tool of the new movement emerged clearly from this first experiment. They were: (1) the limited size of the study group; (2) study and discussion oriented towards intended action; and (3) mutual self-help in study and action, the stronger in mind and resources assisting the weaker.

The idea behind the movement has been well stated in Dr. Coady's own presentation of the story, *Masters of Their Own Destiny*, a book of mellow cooperative wisdom and a personal document of a forceful but genuinely democratic leader of men. It is a statement worth quoting in full: "The Extension Department," says Dr. Coady, "has followed the assumption that every ordinary man or woman is a potential student and every small group of people a potential study club. It presumes that once the people have learned to solve their most pressing problems, they will have tasted the delicious fruit of self-accomplishment which will spur them to the solution of all other problems of life. It is sufficiently realistic to know that 'not alone by bread does man live,' but certainly not without bread. It follows the psychological principle that education to be effective must have a specific objective and that it must be related to the situation which confronts the learner at the time of study. Furthermore, it is based on the educational principle that we learn by doing and by doing the things that bother us, or whose solution needs attention. It presupposes the sociological doctrine that man is essentially a social being, that he finds his best expression in the group and that cooperative study paves the way for cooperative living." [1]

But it is one thing to see all this clearly, and another to make others accept it and act accordingly. The first phase of the movement was by no means one of quick and easy success. The very conditions which called for remedy had bred in the people attitudes which resisted any attempt towards change. To men who had to pour their last ounce of strength into work that, even so, would hardly yield a bare subsistence the idea of taking on the additional strain of study could not at first seem attractive. They were inclined to react with the traditional stereotype of resignation, to prefer to "bear those ills" they had than fly to others that they knew not of, and to argue that old dogs cannot be taught new tricks. In spite of all the efforts, many of the first groups, hopefully started, disbanded after a short time, and left the organisers facing frustration.

The workers in the field were few, but they were determined, persistent, and prepared to take the people as they were, and not as they wanted them to be. They learned from failures, and so they succeeded in stirring some of the people out of their inertia. Results, evident in terms of economic betterment, stronger morale, and a fuller sense of life, followed and gave strength to the movement. It spread from the area originally covered by the Extension Department, the seven eastern counties of Nova Scotia, to the rest of the province, and from here to the neighbouring provinces, Prince Edward Island and New Brunswick. The effects can be gleaned from these simple statistics: The total population of the three provinces is 1,225,500. Of these, about 45 per cent. are engaged in mixed farming; 40,000 are fishermen, and there are 13,000 coal miners and

[1] M. M. Coady, *Masters of Their Own Destiny* (New York; Harpers, 1939, p. 65).

7,000 steel workers in Nova Scotia. These people have organized today about 150 cooperative stores, with an annual turnover of $9 million. The wholesales did a business of $6 million in 1948. The assets of the 451 credit unions amount to $8 million, and their membership to 70,000. The fishermen cooperatives, affiliated in the United Maritime Fishermen, handled in 1947 $2 million worth of business, and the unaffiliated fishermen cooperatives added another million to the total turnover. The cooperative hospitalization scheme has attracted 253,000 members; and there are seven housing cooperatives which have built 85 houses for their members.[2]

But the achievements mean even more than the figures indicate. This, chiefly because the movement has early realized the significance of "multiple cooperatives." The idea itself is not new; it has been proclaimed before, with particular emphasis by George Russell who, in his *National Being,* argued that it was not enough to organize farmers in a district "for one purpose only," but that it was necessary to teach them to absorb "all rural business into their organization," if a real change in the attitude of the people was to be effected, and "true social organisms" were to be created. With this Dr. Coady's own ideas were in agreement. He, too, saw that "the rural people's problem will be solved when the rural people in each large natural economic region of the country will themselves own and operate a complete unit of economic institutions to take care of their own business." He, however, was not satisfied in having made his point. He proceeded to make the people see it and act upon it. The most important single implement utilized in pursuing this aim was the study group.

The method of organizing and conducting these study groups, as it developed and it being effectively practised today, can be briefly described as follows:—

[2] These figures are from M. M. Coady, "Co-operation in the Antigonish Movement." In: H. F. Infield & J. Maier (eds.) *Cooperative Group Living,* (New York: Henry Koosis & Co., 1950).

The first step taken is usually a mass meeting. The meeting serves to stir up the desire for self-help in as many people as can be reached. At such meetings, an analysis of problems concerning the people in the area is presented, and ways and means by which they can be solved are generally indicated. The need for intelligent action which has to be preceded by a thorough preparation of minds, is stressed. Finally, an appeal is made for a voluntary organization of study groups.

If the mass meeting has been successful, the call is heeded by a sizable number of people, and the groups are formed. These normally consist of adults of both sexes, and meet, usually once a week, at one of their members' home, or in a room of a public building, a school, a community center, or a church.

In conducting the meeting, each group is entirely on its own. At its first session, it elects a chairman and a secretary. The main function of the chairman is to prevent the discussion from becoming "blocked," either by persons "hogging" it, or by fixation to one side of the issue. The chairman has to see to it that each member takes his due part in the proceedings, and that the subject is considered from all angles. The task of the secretary is to record the gist of the meeting, and to present it at the beginning of the next, so as to ensure continuity from one session to the next.

The content of the discussions is determined by the line of action the group would like to take in order to satisfy some of their needs. In the beginning of the movement appropriate study material was scarce. The skill of preparing well-organized and clearly-written pamphlets had yet to be developed. Today, the movement has produced a sizable amount of printed and mimeographed material, covering practically every phase of economic life in the area, its history, and culture. A few of the titles of pamphlets published by the Extension Department under the heading *For Study and Action*, may convey an idea of the range of subjects. Some are general in scope, like "Economic Democracy—Its Meaning and Necessity"; "The Key to Progress" (meaning physical, mental, and social fitness, as well as group action); "Scientific Knowledge and Planning," and so on. Other pamphlets discuss the concrete ways and means of cooperative action, such as "Consumer Cooperation—Fields of Action," "Producers' Cooperatives—Marketing and Trade Unions," "Cooperative Medicine," "Automobile Insurance," and so on. More directly bearing on the problems of the farmer is such study material as: "Soil," "Cropping System," "The Compost Heap," "Aids to Long-Time Farm Planning," "Feeding and Care of Cows."

The material, virtually all prepared by the staff and affiliates of the Extension Department, is concisely written, and factual. Before appearing in print, it is often used for presentation on the radio, the CJFX station, owned by alumni of the St. F.X. College. This, in recent years, has become one of the additional media of communication available to the department. In employing this medium, the group technique is by no means being neglected. Local leaders and Extension workers organize listening groups who receive, in advance, written copies of the material to be discussed. The program itself is presented not as a lecture, but in the more dramatic form of a group discussion by a panel of four experts. The subjects center on agriculture, but range as far as health, music appreciation and poetry.

In addition to radio, the film is being used to help people in one area to visualise achievements in other parts of the country. Here, technicolored slides, interpreted by a narrator, prove effective in mass meetings, and in larger gatherings of study groups.

Thus, from hesitant and humble beginnings, a movement has grown in the Maritime Provinces which can claim credit for some sound achievements. It has by no means led the masses from "rags to riches," or replaced hovels with palaces, nor did it ever aim at such futile miracles. What it could do, and has done, is to clear away some of the worst aspects of the rural slums in the area. In the process it has transformed an outlook of resignation and despair into a new "time perspective" of strength and hope. It has given the farmer and fisherman a new morale which imbues him with a sense of vigor and a feeling that, if he wants it, he can become capable of forcing a reluctant physical environment to yield some of life's most precious satisfactions.

Saskatchewan

Another, possibly more advanced use is made of the study group—at least in terms of intended action—in another part of Canada, the prairie province of Saskatchewan. This is true, in particular, of those study groups which were formed spontaneously in this province among farmers who sought a solution to some of their most urgent problems in cooperative farming.[3] Here, the object of the study was agricultural cooperation of a type more comprehensive than that contemplated, so far, by the farmers of the Maritimes.

The difference in the degree of intended cooperative action is not the only one which distinguishes the two developments. The Saskatchewan farmers did not wait for outside stimulation. They rather aroused themselves to the possibilities of cooperative farming. In this sense theirs is a true "grass root" movement. After they had started, they were glad to receive educational assistance. But again, it did not, as in the Maritimes, come so much from an institution of higher learning as from the Government.

There are several reasons why the Saskatchewan farmers should be inclined to think of cooperative farming as a solution to some of their basic problems. One is given in the

[3] "Saskatchewan: A Sound Pattern of Growth," in this volume.

nature of farming in a prairie province. Agriculture consists here chiefly of wheat farming which to be rewarding requires substantial acreage and large-scale machinery. Second, the farms, being large, place the farmhouses at great distances from each other. This makes social intercourse difficult if not, especially in the winter time, virtually impossible. Finally, there is the problem of succession on a farm which if it is to yield subsistence, cannot be left to more than one of the several sons.

Cooperative farming, however, to become acceptable as a way to satisfy any or all of these needs requires a highly cooperative "atmosphere." This, too, could be found in Saskatchewan which among all the Canadian provinces can boast of the densest network of cooperatives. The most powerful among these cooperatives, the Wheat Pool, has for many years now—first in connection with the Extension Service of the University of Saskatchewan, and in more recent years assisted by government agencies— carried on a program of agricultural education. This program culminates each year in socalled "cooperative schools" which assemble rural youth in the towns of the province for one week sessions. The topics treated range, as usual on such occasions, from philosophy of cooperation, through opportunities for youth in agriculture, to cooperative marketing, credit unions and nutrition. As in the last few years, the specific Saskatchewan note was sounded in discussions devoted to the Department of Cooperation and to cooperative farming.

As can be seen, both the general conditions as well as the "cultural atmosphere" have been favorable in Saskatchewan to a trend towards cooperative farming. The farmers, as already mentioned, proceeded to consider action of this kind without any outside prompting. But they were fortunate enough to find themselves encouraged by a sympathetic government. The Cooperative Commonwealth Federation, Canada's Third Party immediately upon coming to power in Saskatchewan in 1944, established the Department of Cooperation and Cooperative Development with a Minister of cabinet rank at its head. This departmen is still the only one of its kind on the North American continent. Among the higher officials of the department has been, virtually from the beginning, a specialist in cooperative farming who, together with his assistants, has since devoted all his attention to the cooperative farms planned or actually formed by the farmers themselves.

Only in one case has the Government itself taken the initiative in the establishment of such farms. In the case of the veterans, returning home from different parts of the world and coming from different units of the armed forces, it was necessary to provide a common meeting ground for those interested in settling on cooperative farms. This the Department of Cooperation did, in collaboration with the Department of Reconstruction. In all other cases its function consisted chiefly in providing factual information for the study groups, in legal advice for the drawing up of the by-laws, and in expert assistance for the setting up of a sound accounting system.

In preparing study material for the groups, the Department of Cooperation was able to enlist the support of an alert and active Division for Adult Education, of the Department of Education. One of the most successful products of this joint effort is a series of 12 study bulletins on cooperative farming. This series of bulletins offers what is probably the most effective step by step presentation of the subject available to-day. The series is opened with an introduction on how to form study groups, and on cooperative principles in general. There follows (in No. 2 and 3) a brief sketch of the history of farming in the province, and a discussion on how to apply the Rochdale Principles to cooperative farming. The following issues deal with pertinent instances of cooperative farm practice. Three of the issues (Nos. 4, 5 and 6) describe the cooperative farms existing in the provinces. Each of the following three issues (Nos.7, 8 and 9) is devoted to significant developments abroad, the *kvutza,* the *kolkhoz,* the F.S.A. cooperative farms, and the collective *Ejidos.* Brief mention is made of the Hutterites, and some cooperative communities of the past. The last two issues (Nos. 11 and 12) discuss procedures to be follwed when actually organizing machine-sharing cooperatives and complete cooperative farms.

In addition to factual material, the bulletins contain bibliographical notes, reference to source material, interspersed with advice on how to conduct the discussion to the greatest benefit of the members. Some of these "Tips on Group Study" might be worth quoting. There is, for instance, one that calls attention to the importance of physical arrangement: "Before opening the meeting, see that everyone is comfortably seated, preferably in a circle or semi-circle, that lighting and ventilation are good and that all members are introduced to one another." Or, another one, which admonishes: "Keep your meetings friendly and informal. Complete agreement on every question is not important. What does matter is that every person should feel he (or she) has contributed, and has learned something from discussion." And still another points out: "Each question should be carefully considered before going on to the next. Don't worry if you find it difficult to get through all the questions in one evening. One or two questions thoughtfully discussed are more valuable than three or four questions glossed over."

According to latest information study groups considering cooperative farming are spreading throughout the province. The actual number of cooperative farms is as yet small. The first of these farms was founded in 1944; by 1947 nine had come into existence; at latest count in December 1952, there were 27, formed partly by farmers in the different parts of the province, and partly by veterans on Crown land.

Indicative of the slow but steady growth of the development is the fact that already in 1947 delegates of the groups met and decided to establish a federation of all cooperative farms in the province. An even more revealing sign of the development's potential strength might be found in the circumstances under which it is taking place. The Saskatchewan farmer enjoys at present his full share of North American agriculture's prosperity. Once times, as they have done before, become less propitious again, will not the urgency of needs become common to all? Will not then the solution offered by cooperative farming appeal to a number of members far surpassing to-day's small pioneering group of the most far-sighted?

The Pioneer Training Farms

The *Kvutza*, the modern cooperative community which came into existence in connection with the Zionist resettlement enterprise in Palestine, practises what is probably the most comprehensive kind of cooperation.[4] Since its origin, shortly before the outbreak of World War I, this kind of cooperative living more than any other aspect of the development in Palestine attracted the attention of Zionist youth all over the world. The intention to go to Palestine became for many identical with that of joining the *Kvutza*. Preparation for this step meant for a youth of predominantly urban background, learning to adjust to a change incomparably more incisive than that implied in the "multiple cooperatives" of the Maritimes, or even the cooperative farms of Saskatchewan. Jewish youth had not only to learn to do things in a new way, but to learn to do things they had never done before, and to learn to do them in a new country, under strange and most difficult geographic and climatic conditions. To prepare their minds, however thoroughly, was not enough; physical training

[4] "Sociology and the Modern Cooperative Community" and "Testing of a Pioneer Training Farm" in this volume.

became just as, if not more, important. Learning, unaccompanied by doing, had little effect. The study group was developed to a point of virtual identity with the situation for which it prepared. The *Pioneer Training Farm* is in every aspect of its socio-economic organization a replica of the *Kvutza*.

The agency in charge of this development is known by its Hebrew name, *Hekhalutz*, which means "The Pioneer." It was founded, shortly after the end of World War I, in Europe. It set itself as its main goal to offer guidance to Zionist youth in their effort to prepare themselves for emigration to Palestine and especially for life and work in the *Kvutza*. Concentrating at first on the countries of eastern Europe, the organization quickly spread to the rest of the continent, and later to other parts of the world. By 1939 it had established branches in every country on the globe where a sizable Jewish population was found. During World War II, when the fury of Nazi extermination struck European Jewry, the Pioneer Organization was forced to abandon its educational activities. It turned into a militant underground formation, and provided some of the outstanding leaders for the Jewish resistance, including the military organizers of such heroic uprisings as that of the Battle of Warsaw. Today, with European Jewry virtually gone, and with the Jews of the Soviet Union beyond reach, the Jewish population on the North American continent remains the last stronghold of Zionist aspirations. It is here that the Pioneer Organization, supported by the General Federation of the Workers of Israel *(Histadrut)* and the Jewish Agency, is concentrating its best efforts.

When the Pioneer Organization first began to carry out its program in eastern Europe it found itself handicapped not only by the scarcity of funds, but also by legal barriers. In most of these countries Jews were not allowed to own farms. To acquire farming skill, Zionist youth had to hire themselves out as farm hands. This they did wherever possible in groups, or by taking jobs in the same area and trying to keep in touch with each other in their spare time.

No such difficulties are encountered in the U.S.A. or in Canada. Here anybody may buy a farm if he has the money. But other factors tend to hamper the efforts of the pioneer. Chief among them are American Jewry's general situation, and the Zionist Organization's own attitude.

There is much that could be said about the difference in the situation of American Jewry as compared with that of Europe. What interests us here is that aspect which has a bearing on the work of the pioneer. An indispensable premise of this work is a desire on the part of Jewish youth to emigrate to Israel. In America, conditions do not exert a strong pressure in this direction. True enough, there is anti-semitism and discrimination. But anti-semitism here is mainly of the "social" kind. Vicious as this variant certainly can prove to be, it tends to affect the Jews' sensibilities rather than their livelihood. It will induce, therefore, a more or less localized resentment rather than a will to emigrate. The wealthy Jews, if they set their mind to it, can easily out-snob their adversaries—as illustrated by the real-estate man who, excluded from a hotel because of restrictions against Jews, forced out the owner himself by buying the place. The resentment of those in more modest circumstances, to say nothing about the far more numerous poor Jews, is much less militant. Their struggle for life is much too exacting to leave much room for social ambitions. What affects these strata more acutely is discrimination, in higher education, in the professions, and in employment. So long, however, as there are jobs for everyone there is no need for emigration. Should unemployment become wide-spread again, the "interpretation of the situation" might well change.

Probably related to this situation is the attitude within the Zionist Organization itself. The numerically and financially strongest section of the American Zionists, the so-called General Zionists, has until quite recently shown very little concern for the aspirations of the labor sector of Israel. The efforts of the Pioneer had to depend mainly on the support of the much weaker Labor Zionists. Little wonder, then, that the growth of the Training Farms has been slow and uneven. For many years only one farm existed in the U.S.A., to which a second was added shortly before the outbreak of World War II. Both were located in New Jersey. Even with all the enthusiasm generated by the history-making events that led to the establishment of the State of Israel, their number never rose to more than eight. In November 1951, the Executive Committee of the Labor Zionist Youth (*Habonim*) announced a change in its policy which implied the closing of the three Training Farms under its jurisdiction and the establishment of a Youth Workshop in Israel instead. Subsequently, several of the other farms were also liquidated, thus reducing the number of those still in existence at present to two again.[5] This is certainly a more than modest showing for a Jewish population of some five millions.

Few as they are in numbers, these Training Farms are of great interest to the American student of education for agricultural cooperation. They offer an opportunity to observe directly one of the many striking aspects of the *kvutza,* the preparation it has organized for prospective members. All of these farms are run exactly like a *kvutza,* with one obvious exception that their purpose is training and not permanent settlement. Membership is limited to the training period of, normally, one year. Being educational institutions, they are directed and supervised by instructors and agricultural experts, usually emissaries from Israel.

Admission to the farms is open to Jewish youth over 18, of both sexes, who are members of the Pioneer Organization and

[5] Cf. this author's "The Closing of the Zionist Pioneer Training Farm in America," *Cooperative Living,* III, 3 (Spring, 1952).

who have been recommended for agricultural training by one of the city study groups of the organization. As a member of the farm, each trainee has a vote in the General Assembly, which is the highest authority of the group. He is also eligible for any of the committees which are in charge of the economic, social, cultural, and recreational activities on the farm. The farming practised is of the mixed type and tends to include all branches that can be cultivated in the given areas. Each member has to take his turn in "unproductive" work, kitchen duty, general cleaning, etc. According to the regulations of the Pioneer Organization, the only finances required for joining are funds for the passage to Israel. All other expenses are covered by the Organization, but all money earned on the farm or in outside work goes into a common purse. Community of goods extends to such personal possessions as work-clothes, books, radio, records, and so on.

A majority of the members come from lower middle-class urban families. To adapt themselves to life and work on the cooperatively run Training Farm means to all of them an almost complete departure from their previous ways of life. It means, as one member put it, "to change the stream of their life to become in many cases the first real workers of their families in generations." It means also adapting oneself to a radically new kind of social existence. To quote from the same source again[6]: "The conception of human relationships . . . is very different from that of the ordinary society." For it implies among other things: "The abolishing of the money relationship, the real equality of the sexes, the conception of working not merely for individual gain but for a group, the fact that the respect of others is gained not by the amount of property you own, or the physical strength you have, but by the per-

[6] Michael Cohen, "Hakhshara (preparation) for What?" *Palestine Information* (September 1947).

sonality you have." It implies, in short, a new type of society "with a new relationship between men."

The strain which this radical readjustment puts upon the individual cannot but be extraordinary. Inner and outer conflicts must and do arise. The outcome depends, as in group formation generally, on the crystallization of a strong and resistant nucleus. The fact that virtually all of those who underwent training on the farms formed group "kernels" who emigrated together, seems to indicate that the Training Farms succeed in creating a strong we-feeling among their members. This does not mean that the farms differ from other educational institutions in that they do not know "flunking." They, too, have had their share of those who "cannot take it" and quit, and of others who, even though they graduate, cannot make a go of it in Israel and return. Failures of this kind are hardly avoidable. They suggest, however, that the selection of trainees and the methods of training could stand some improvement.[7] All the same, there is the undeniable fact that the American trainees have made a significant contribution to the *kvutza* movement. Many have become highly appreciated members of existing *kvutzot;* others have taken the initiative and an active part in the establishment of new settlements. There are today six such settlements which are commonly referred to as "American": *Ein Hashofet, Kfar Blum, Kfar Menahem, Khatzor, Mayan Barukh,* and *Ein Dor.*

Applications

Proceeding from the observation that cooperation, to become more than a matter of mere economic convenience, needs re-education, we have presented three instances of education for agricultural cooperation. We found that re-education

[7] Research undertaken in the American Training Farms by this writer was designed to gather the facts on which such improvement could be based. Cf. "Testing of a Pioneer Training Farm" in this volume.

is being practiced with apparent success in the Antigonish Movement, in the cooperative farm groups of Saskatchewan, and in the Training Farms of the Zionist Pioneer Organization. In each of these instances the most effective technique employed is that of the study group. Essentially, this technique is in all three cases the same; it varies only, depending on the character of intended cooperative action, as to the degree of activation.

The immediate occasion for this particular discussion is the fact that the governments of several countries are considering cooperative farming as a means for the solution of some of their urgent rural problems. Successful cooperative farming depends, like all cooperation, on the voluntary and active participation of those who are to benefit from it. To elicit such participation, no other means can be as effective as education. The instances cited above may prove to be useful models.

Now a few words only on how the lessons learned might be applied. The application will be all the more fruitful if due attention is paid to two main and closely allied aspects of the task: (1) the essential nature of the study group, and (2) the basic requirements of the intended cooperative action.

The significance of the study group lies in the fact that it contains socio-psychological features which make for genuine cooperation. If it is true, as the late Kurt Lewin held, that the chances for re-education seem to be increased whenever a strong we-feeling is created," [8] then it is the group rather than the study aspect that is of first importance. It is not so much what is being learned that counts as how it is done, the active participation in the common process of learning. Each member is as much teacher as he is pupil. He joins the group out of his own free will. He thus chooses freely the people with whom he studies. He has an equal voice in all the proceedings of the

[8] Kurt Lewin, Resolving Social Conflicts (New York: Harpers, 1948), p. 67.

group, and it is up to him as much as to anybody else to select the subject to be studied. If any outside leadership is present at the time of the formation of the group, it is restricted to the offering of choice alternatives merely; the actual decision is left to the group itself. Its conduct is determined by the needs of the members. Authentic leadership develops from the ranks of the membership. Any authority will be acceptable so long as it enhances the functioning of the group, it will remain effective so long as it acts in this direction. In short, in every sense of the term, the study group is a democratic body.[9]

The processes underlying cooperation, particularly where, as in cooperative farming, cooperation aims at becoming a way of life, are essentially the same as in the study group; they amount to democracy in operation. The study group, where care is taken to heed the factors which make for "co-operational" democracy, is, therefore, likely to achieve the best educational results. Among these factors the following seem to be of particular importance: (1) capacity for action; (2) democratic leadership; (3) the needs of the participants.

(1) *Capacity for action.* Cooperative farming is often resorted to because all other attempts at remedying the plight of a farming population have proved futile. This implies that the ills to be cured are of long standing, and that they probably include all the evils which characterize sub-standard existence, undernourishment, disease, illiteracy, and mental narrowness. Conditions of this kind tend to reduce people to resignation and apathy. Now cooperation means above all *action.* Before people can become capable of any, the very conditions of their existence must be changed. Hence, cooperative farming must be preceded by a comprehensive program of physical and mental rehabilitation.[10] If in carrying out such

---

[9] *Loc. cit.,* p. 76, and *passim.*
[10] Cf. "Health and Agriculture." In: *Yearbook of Agricultural Co-operation* (London: The Horace Plunkett Foundation, 1947).

program care is taken at every step to enlist the active participation of the people, much will have been done to prepare the stage for successful cooperation.

(2) *Democratic leadership.* The task of introducing cooperative farming will be easier where at least part of the population enjoys somewhat higher standards of living. The natural tendency of an administration will be to concentrate its efforts on the more advanced sector of the population. This approach will probably produce a quicker response and prompter action. Care, however, must be taken that those encouraged to assume leadership become fully imbued with the meaning of cooperative action. For although it is true that capacity for reform is far less common than that to conform, passive acceptance contradicts the very essence of cooperation which, more than anything else, means *doing* things together. The lead may well be taken by the more energetic and far-sighted, but caution must be exercised that their advance be kept within a pace which will not bring them out of step with other peoples' capacity for active participation. Only then will leadership avoid the pitfalls of autocracy which, even if benevolent and paternalistic, will always remain inimical to genuine and successful cooperation.

(3) *The needs of the people.* The study group, as we have seen, is quite adaptable. It can be used with the same effectiveness by such different people as the farmers and fishermen of Canada and the urban Zionist youth in many parts of the world. In utilizing this technique, though, it must always be remembered that in our world today cooperation, as we have pointed out, involves a departure from prevailing norms of behavior, a more or less drastic change in peoples' attitudes and habits, depending on the inclusiveness of cooperative action. For cooperative farming, the extent of required change will be considerable. A change that is to be of more than merely transitory duration can be effected only if it is prompted

by some real and urgent needs of the people. The basic needs of people, for food, clothes, and shelter, are virtually identical in all parts of the world. But the means and modes of their satisfaction differ almost infinitely from one often relatively small cultural area to the next. In such matters as dress, diet, and other personal tastes, they may even vary from one village to another in the same neighborhood. The mechanical application of remedies which proved successful in one place, may therefore meet with resistance and resentment in another. Such opposition can be reduced to a minimum if the proposed reform takes full account of people's culturally conditioned modes of satisfaction as well as of their ascertained needs. This will apply doubly for the study group and for cooperative farming.

The successful handling of the above factors demands in each case a great deal of specialized knowledge. Such knowledge lies in the domain of the social sciences and can be offered, in particular, by such disciplines as human ecology, sociology, social psychology, and cultural anthropology. It is doubtful whether, under ordinary circumstances, there will be administrators who have mastered at best more than the elements of any of these disciplines. Nor should such mastery be justly expected of them. Depending on the scope of the program, they will need the assistance of a smaller or larger staff of active and well-informed social scientists. More than any other measure, provision for such assistance may prove decisive in any attempt to introduce cooperative farming.

(9)

THE URBAN COOPERATIVE COMMUNITY

The Issue

As yet hardly noticed in America, a development has been
taking place in Western Europe, mainly in France, that should
be of considerable interest to students of community, particularly those concerned with modern cooperative communities.
According to the lastest count, some 1,500 workers' families
have started, since the end of World War II, more than eighty
projects designed to lead to the establishment of what they
designate as *Communities of Work*. A first full report on this
development was given in *All Things Common* by Claire
Huchet Bishop.[1] The author visited some thirty of these communities in the spring of 1950. As she presents the facts, these
communities may be developing the first successful attempt
at solving the somewhat puzzling issue of the urban cooperative community.

To grasp the full significance of this undertaking, we
should recall that, probably because historically it arose in
connection with resettlement and agricultural reform, it is
land settlement that has been generally assumed to be the
necessary condition of modern cooperative community. However, the concept of cooperative community does in itself
not imply any specific kind of economic basis. It refers
rather to any group of people deliberately practising cooperation in their social as well as their work relations. The fact
that virtually all modern cooperative communities are mainly
engaged in farming merely indicates that agriculture and rural

[1] New York: Harpers, 1950.

180

conditions have thus far been more favorable to the development of cooperative communities than industry and urban conditions.

In many respects cooperative communities exemplify a novel way of life. They naturally followed a path of least resistance when they sought out sparsely populated places and the more community-minded rural areas to establish themselves. Once they succeeded as land settlements, the question was bound to arise whether cooperative communities were necessarily confined to the countryside, or whether their principles could be applied to urban areas as well. This is a crucial question in any attempt to estimate future possibilities of cooperative development. True, rural reform cannot be considered a matter of minor importance so long as in many parts of the world large numbers of people are compelled to derive their livelihood from a kind of agriculture that is unable for longer or shorter periods to provide them with even so much as bare subsistence. To the extent that cooperative farming tends to increase food production, it has an important role to play in the rehabilitation of so-called backward areas. It cannot be denied, however, that today the key to the world's economy is held not by agriculture but by industry. Cooperative farming itself depends for its success largely on progressive mechanization of production.

Industrial production and technological progress, as developed to date, have gone hand in hand with crowded population centers. Urbanization, culminating in the large city, the metropolis, has been a dominat trend of industrial society. As has been pointed out frequently, the effect of urbanization has been to compartmentalize all human activities, to split man's thinking and acting into numerous but essentially disconnected pursuits, and thus to impair his sense of belonging and the wholeness of his personality. The "alienation" and impersonality of all human relationships in urban civilization is experienced by large numbers of people as a denial of most essential human satisfactions. Cooperative communities have demonstrated their capacity to yield such satisfactions. Hence the question of whether the cooperative community idea can be applied effectively to urban life is of more than academic interest.

Cooperation and Industry

The thought of applying the cooperative principle to urban occupations, and to industry in particular, is not altogether new. It lies at the very root of the cooperative movement. Robert Owen saw in cooperation mainly a better way of handling the social issues arising from industrialization. He advocated it as an alternative to competition. Owen's solution to the evils of the Industrial Revolution implied a total reshaping of the workers' existence, by means of "villages of cooperation." His valiant but vain attempts at realization of this scheme are today of interest mainly to the historian of cooperative communities. They never had any real effect upon the condition of city laborers.

Of far greater consequence was the workers' own initiative in organizing producers' cooperatives. The development of the Cooperative Workshop, or the Industrial Cooperative, runs parallel with that of the consumers' cooperatives. It is, to be sure, far less impressive, but certainly not as disappointing as is widely assumed. Even in Great Britain, where they met with little favor, there are some ninety industrial cooperative societies at present. Their membership is estimated at 18,000, and their trade amounts to about ten million pounds a year. Their production is limited to virtually three trades: clothing, footwear and printing.[2]

[2] See Margaret Digby, *The World Cooperative Movement* (London: Hutchinson's University Library, 1948), pp. 57 ff.

More extensive is the movement in France, considered by some "the homeland" of industrial cooperation. According to recent releases of the Confederation of the Workers Cooperative Productive Societies, there are close to seven hundred such societies in existence in that country. They are said to employ more than 35,000 workers, of whom less than half are members. Approximately 400 of them carry on building construction and public works. Other trades represented are printing, metal works, timber industries, textile and clothing, quarrying, transport, films, glass works, and others. It is worth noting that a few of these societies anticipate in some of their features the *communities of work,* for example, *The Makers of Precision Instruments,* and the *Familistère of Guise.* The former, founded some fifty years ago, are housed in a large, modern factory building in Paris. They employ over 1,000 men and women, and their annual output of telephone equipment and instruments for ships and airplanes is substantial. A characteristic feature of this society was its simple scale of remuneration: each member, whether director or ordinary worker, received the same wages. This system was replaced by a differential wage scale only a few years ago. The Godin stove foundry, widely known by the name of *Familistère of Guise,* occupies a unique place in the development of French industrial cooperation. It was founded as a private enterprise by Jean-Baptiste-André Godin who, as an ardent follower of Fourier, began to convert it, around the middle of the nineteenth century, into a Phalanstery. When Godin died in 1888, he left the foundry to his employees. Since then it has been run by the workers themselves. The employees are divided into several classes, of whom only the first is entitled to vote in the General Assembly. The workers and their families live in the so-called Familistère, where each family has its own quarters and shares with others in the general services. The foundry was destroyed during World War I, but was re-built shortly after its end and has continued to prosper.

The industrial cooperatives that almost two decades ago attracted the widest attention were those organized by Chinese workers in the wake of Japanese invasion. In 1941, there were 1,867 societies, with a membership of about 30,000. Their activities included machine and metal works, mining, textile, tailoring, chemicals, foodstuff, stationery supplies, carpentry, masonry, and transportation.[3]

City and Community

The rapid break-up of most of the Chinese industrial co-operatives occurred for reasons making for the high turnover of production societies in general. Apart from the confused political situation on the Chinese mainland, these reasons are partly of an external, partly of an internal nature. One of the main external problems into which societies of this kind run is that of a stable market. In England, this problem is solved chiefly through an arrangement with the consumers' cooperatives; in France, through the instrumentality of helpful authorities. But such dependence on outside goodwill is at best precarious. Internally, the main difficulty of industrial cooperatives arises from what might be called a lag between working and social relations. Participation in a production cooperative must be far more intensive than that in a consumers' society, a credit union, or a marketing cooperative. It means not only sharing all of one's working hours with others. It implies also the joint making of plans, of arriving at decisions, of sharing financial troubles and successes. All these are vital matters. For a working man they are among the most crucial in his life.

[3] Cf. "The Case of the Chinese Industrial Cooperatives" in this volume.

All the same, the level of cooperation is equal to that in the consumers' store. All relations are limited to economic transactions, and other matters are left outside. In other words, there is a basic discrepancy between the actual sphere and degree of contact, which is comprehensive, and the nature of the cooperation practiced, which remains segmental. The only possible way of bridging the gulf is to extend the organization so as to include social relations as well.

By becoming communities, it would seem, production societies could solve both their external and internal problems. They become their own market by producing primarily for self-subsistence; and they harmonize their work and social relations by shaping both in accordance with the same, that is, cooperative principles. Obvious as this idea may seem, it is easier to prescribe than to effectuate. The growth of a community is a complex and subtle process which depends as much on spontaneity as upon planning. A community is a blend of contradictory traits. It must be anchored in the minds, attitudes and emotions, and depends on the full participation of all concerned. Such participation must not be forced, but remain free and voluntary. The vitality of a community depends mainly on the intensity of we-feeling it engenders. Yet it cannot thrive on an unsound economy. To avoid the pitfalls of isolation, a community must keep in step with the world around it; to preserve its identity, it must know how to shield itself against adverse influences. In short, it must succeed in being at the same time a world within itself and a part of the world as a whole. This might help explain why the growth of communities is so much more difficult in the city than in the countryside, and why apparently it required the double shock administered to their economy by two world wars as well as a great deal of ingenuity on the part of some pioneering minds among the French workers, to produce what seems to be the first effective approach to the urban cooperative community.

The City Kibbutz Ef'al

Before this approach produced results substantial enough to warrant wider attention, there was on record only one instance of an attempt at forming an urban cooperative community. This attempt was undertaken in the country of classical community experimentation, in what is now Israel. There, in a suburb of Tel-Aviv, some eighty working men and women established in 1947 a "city *kibbutz*", which they called *Kibbutz Ef'al,* with the aim of applying to the life in town the same principles of the *kibbutzim* that had proved so successful in agricultural settlements. With the help of national agencies, they built themselves homes, altogether 72 flats, arranged in such a way as to make possible a *kibbutz*-like handling of all their affairs. They differed from the normal *kibbutz* in that they not only did not work together but depended for their livelihood on different manual and office jobs in the nearby city. By 1948, the membership of this city-*kibbutz* included 106 adults and 65 children. Forty-three of the members spent their days at their employment in Tel-Aviv. Their wages and salaries went into a common purse from which, again as in a *kibbutz,* each received his share, not according to the amount he had earned, but according to his needs. By pooling their resources they achieved a great deal of economy. Their social life became richer, and they felt that they could enjoy a deeper sense of economic and psychological security than otherwise.[4]

However, *Kibbutz Ef'al* remained an isolated instance. The experiment did not seem successful and attractive enough for others to emulate it. When last heard of, it was said to have run into difficulties connected with the differences in salaries and wages earned by the members. Those whose income from their city jobs was higher apparently were not altogether happy about having to subsidize the less skillful and the less

[4] Cf. "Kibbutz Ef'al—City Kibbutz: A New Social Experiment," *Palestine Information* (February-March 1948).

hard-working. Whatever the reason, *Kibbutz Ef'al* is still the only one of its kind in Israel, and leaves the problem of the urban cooperative community still open.

A Dutch Project

The first news of a new and more pointed attack on this problem to reach the GFRI came in a communication from Holland. According to H. J. Van Steenis, general manager, the *Cooperative Vereniging Ingenieursbureau Van Steenis W.A.* in Utrecht aims at "the promotion of the material interests of its members, as well as the serving of human society by the joint labor of these members." The special business of the association is "the execution of architectural and hydraulic engineering and surveyor's operations in the widest sense of the word, both at home and abroad." The staff, consisting in part of members and in part of "co-workers" who may become members after a minimum of one year of service with the bureau, receive the normal salaries. Profits, however, are distributed in a cooperative though somewhat unusual way. Twenty per cent goes to the staff, and sixty per cent to a "Wage Stabilization Fund" which assures the members a steady level of income, even when business is bad. The remaining twenty per cent is returned to the clients, in order to restore to them—as the statutes explain—the overcharge revealed by the accruing of profits.

The other points of the association's statutes are more or less consistent with the cooperative nature of the enterprise. The staff now numbers fifty, of whom twenty-six are members. The association carries on projects not only in Holland but also in Turkey, Indonesia and Africa, and its finances are in a very satisfactory state. The working conditions are those common in such enterprises, although slightly more favorable to the staff. Thus, for instance, members who want to attend one of the Folk-High-Schools, of which there are some ten in

Holland, receive an extra week of vacation. Most interesting, however, is a passage in the December 1950 issue of *Communauté,* the organ of the French communitarian movement, which relates that Mr. Steenis, who founded the Bureau in 1945, "from the beginning intended to turn the enterprise into a *communitarian* one." This intention, we are told, Mr. Steenis was able to carry out in 1948, without benefit of precedent, "the experience of the French communities being unknown to him at that time."

Marcel Barbu and the Beginning

As we were only vaguely aware of the existence of a French watch-case factory claiming to be organized as a community, the above passage appeared somewhat puzzling and intriguing. There was, surely, a story behind those lines, possibly one of great significance to students of group living. That story has now been told, candidly and with the deep understanding that comes from first-hand observation, by Claire Bishop. It might well be, as the flap of her book contends, that the communities she describes have sprung up spontaneously and "almost simultaneously," meaning that they (like the one in Utrecht— which, by the way, does not seem to have reached as yet the status of a full community) conceived the same plan and idea independently. As Mrs. Bishop tells the story, there can be hardly any doubt, however, that they all learned from, and modeled themselves after, the longest-lived and apparently most successful of them all, *Boimondau.* It may well be considered the prototype of the *community of work,* a kind of *Dagania Aleph* of French communitarianism. Its present name is derived from the first parts of the three French words *boitier* (case), *montre* (watch), and *Dauphine.* It first called itself *Community of Work Marcel Barbu,* and its origin is closely connected with the man of this name. As Mrs. Bishop relates, the story of Marcel Barbu is that of a self-made man with a social conscience, somewhat resembling Robert Owen in this respect. While Owen, however, kept pleading and associating with the "upper classes" to which he had risen, Barbu remained with the workers. He began as a manual worker, his trade that of watch-case maker. He learned to know the hard lot of the French worker from his own experience. He obviously disliked being coerced, whether by his employer or his union. One day he and his wife decided to start a business of their own, not to get rich but, as Mrs. Bishop quotes him as saying, "to shape our own means of liberation." The beginning was, of course, filled with hardship. The two had neither money of their own, nor credit, nor any other kind of backing. In order to buy machines, they sold their furniture. To make a go of it, they sometimes worked twenty hours a day.

Once his enterprise was functioning, Barbu introduced a factory council. He gave the workers a part in vital decisions, such as in matters of determining their own wages. "Labor-management collaboration" of this kind was nothing novel. For Barbu, it signified only the first groping step toward a new form of industrial enterprise he meant to create, but for which he knew no precedent and had no blue-print. He merely felt that a new industrial and social structure would result from a saner relation between man and his work. He knew it could not be established by one man alone, and he asked his workers to participate in the search. Alas, they were not interested. They found conditions in his factory better than elsewhere and were satisfied to accept them as they were. It was only in 1940, when breakdown and defeat made business as usual impossible for all, that Barbu succeeded in rounding up a group of workers who were willing to try new ways. Even so, he had to take the men as he found them—there was among them a barber, a sausage-maker and a waiter—and taught them his trade. He did so on one condition only: that they would help him in his effort to find a way to abolish the distinction between employer and employee.

In Valence, a town of 30,000 in the South-East of France, Barbu and his men leased a barn, and went to work producing watch-cases. Two months later the first orders came in, and the first results of their new set-up began to make themselves felt. The first thing they discovered was the immensely liberating effect of genuinely free speech between employer and employees. So much did they exuberate that their work began to suffer. Disciplining themselves, they relegated all discussion to a regular weekly meeting. This led to the next significant finding. What they were after was the kind of mutuality that would allow for a maximum of freedom and independence. There would have to be, they could see now, a limit here. Too much license could easily lead to disintegration. They felt the need for some kind of fundamental binding force, and realized that it would have to be of a moral nature. They proceeded, then, to formulate what they came to call their "common ethical basis" which contained such "commandments" as: "Thou wilt be faithful to thy promise . . . Thou shalt respect thy neighbor, his person, his liberty . . . Thou shalt respect thyself . . . and Thou shalt hold that there are goods higher than life itself: liberty, human dignity, truth, justice." This was not a rigid code, but a minimum of ethical consent, open to modification and, if possible, to improvement in the light of changing conditions and growing experience. In other words, it was a code of functional ethics. A by-product of their finding was their insistence on unanimity rather than majority in all decisions. For consent based on ethical principles cannot be summed up in fractions; to be sound, it has to be wholehearted. Finally, and as a natural sequence to the two preceding discoveries, there was the realization of the need for more knowledge. That meant education, which required more leisure-time. To gain it, they decided to put more effort into production and to use the hours saved from work for instruction.

After proceeding in this way for two years, they discovered that they already had found what they were searching for, that they had ceased to be just another factory, that they had become a "community of work." The step formally recognizing this actual condition was taken when Barbu, as Mrs. Bishop puts it, turned over to the community "the factory that had been created by the work of all." The act of transfer stipulated that Barbu was to be repaid the sum he had put into the common enterprise and that, in case he wanted to start a new community, he had the right to take certain of the machines with him.

The Community of Work

In this way the first *community of work* came into existence. The men who shaped it were workers. Their motives, whether they were aware of it or not, were essentially the same as those that led to the establishment of the first successful consumers' store in Rochdale. The desire to become "masters of their own destiny" had prompted the Rochdale Pioneers, as it evidently prompted Barbu and his associates. In both cases there was readiness for direct, albeit non-violent, action, that is, cooperation. While the ultimate intentions of the Rochdale Pioneers were frustrated by the success of their store, the French communitarians took up, consciously or not, the Rochdalers' own original objective, the "home-colony of united interests," and they are moving towards it in no devious manner.

As industrial workers, intent on becoming truly independent, the people who started *Boimondau* sought, almost naturally, to regain control over their means of production. This they achieved by resorting to cooperative production not, however, without realizing that they would have to go beyond it if they wanted to avoid the frustrations of the Industrial Cooperative. Their—and the urban cooperative community's—basic problem was how to organize industrial work so as to make it yield the basic human satisfactions of belonging and of personal fulfilment. They solved this problem, or at least showed the way towards its solution, by the ingenious set-up they found for their *community of work*.

Boimondau has been in existence for more than ten years. Its legal description is: "Community of Work, Workers Cooperative Production, Limited, with variable capital and personnel, in communitarian form." Its form of organization, although by no means final, has had time enough to solidify. It may be taken as representative of the goals aimed at by the other communities discussed by Mrs. Bishop. The main features of that organization, as culled from the account given of them in *All Things Common,* may be briefly summarized as follows:

Membership. Populationwise, *Boimondau* is "like any other cross section of French life." During the war, most of the members were in the Maquis. Nine of them were arrested by the Germans, and three were killed. Of the 273 persons that belonged to it at the time of Mrs. Bishop's visit, eighty were children, and fifteen under 21 years of age. 178 of the members were "full-fledged French citizens," and of these, 167 belonged to no political party. As to faith, the community was divided in the following way: 110 Catholics, 70 Materialists, 58 Humanists, 32 Protestants, and 3 Undecided. Mrs. Bishop explains that "Materialists" were those who believe in "matter as the only reality;" "Humanists", on the other hand, "ranged from skeptics to people who did not think they would quite join the Christians." Jews were listed under both Humanists and Materialists, as none were orthodox.

In order to become a full member or "companion," an applicant has to pass through three probationary stages. Upon presenting himself at *Boimondau,* either on recommendation of a member or on his own, he is interviewed by the Chief of Community. The interview is designed to establish his technical as well as his personal qualifications. If he is a skilled worker and otherwise acceptable, a place is found for him appropriate to his skill and experience. If he is new to the trade, he is put through the standard aptitude tests. In any case, he first becomes a salaried worker or apprentice. As such, he participates in all social activities and is expected to become acquainted with all the rules and regulations. After three months, he may ask to be accepted as *postulant.* If his request is declined, he must leave; if not, he enters into what is called "the novitiate." In this stage he enjoys all the duties and rights of full membership but the one of voting in the General Assembly. At the end of a year his qualifications are discussed, thoroughly and from every angle, by all concerned. If unanimously accepted, he becomes a *companion;* if not, his term of novitiate is extended for another period of time. In a sense, even the status of a *companion* is not final. Each member is subject to re-examination of his status, which is carried out by the General Council at regular intervals of three months.

To comply with the law, *companions* are expected to subscribe to a share of the cooperative, which is the *community's* legal definition. Only those working in the community can hold a share, on which, incidentally, no interest is paid. If a community is dissolved, the remaining funds are not shared out but are instead handed over to another community. The *community* distinguishes two kinds of *companions:* A "productive" *companion,* who works in the watch-case factory; and a worker's wife who has passed through the three stages described above, called a "family" *companion,* or one who contributes to the welfare of the community mainly through home-making. Both have the same right of vote in the General Assembly.

Work. The attitude of *Boimondau* towards work is best expressed in the following explicitly stated principles quoted by Mrs. Bishop: "(1) In order to live a man's life one has to enjoy the whole fruit of one's labor. (2) One has to be able to educate oneself. (3) One has to pursue a common endeavor within a professional group proportioned to the stature of man (100 families maximum)[5]. (4) One has to be actively related to the whole world." Accordingly, work is considered mainly as a means for personal fulfilment. Every activity, whether it contributes directly, through industrial production, or indirectly, through self-improvement, to the welfare of the community, rates as work. It may fall into either of the two main sectors, the industrial or the social. The industrial sector, which at *Boimondau* centers on the watch-case factory, is organized in teams, sections, and services. A team consists usually of not more than ten men, a section of several teams, and a service of several sections. Similarly, the social sector is divided into teams and sections, such as the "Spiritual Section", the "Artistic Section", the "Communitarian Life Section", the "Sports Section", and so on. Up to nine working hours a week are devoted to work in the social sector.

Ownership of the means of production (machines, buildings, ground, stock capital) is social, or as Mrs. Bishop puts it, "collective and indivisible." The members "have the use and fruit of it." Consumption, however, is fully individual. Remuneration for work in the industrial as well as the social sector is differential, though based on the same principle. It aims at returning to each the "whole fruit" of his contribution

[5] We shall see that *Boimondau* itself has not strictly adhered to this limit; according to latest counts, its membership had grown to 150 families.

to the community. For example, in the social sector the worker is paid for educating himself, or for getting well when he falls ill. Concerned with doing full justice to each member, *Boimondau* uses a complicated system of points and coefficients in its scale of remunerations. The complexity of this system and the continued efforts to refine it are evidence of the difficulties encountered and the scruples taken to meet them in a spirit of equity. It is possible that the final arrangement will prove quite simple—once the communal spirit will have matured to the point where differential payment, and monetary rewards generally, will have lost its meaning. As things stand at present, equalization of points credited for professional and social contribution makes it possible for one "productive" *companion* to outrate another equally good one, because of superior social performance. It is the precise assessment of the social value of a member, however, that presents the most acute problem. It would seem that knowledge of some of the newer social techniques of rating, such as the sociometric test, could prove of help here.

Each *companion* is entitled to a full month vacation. To become acquainted, by way of personal experience, with the work of her husband, each woman works a whole week in his special factory section. The *community* assigns a person to take care of her household duties during this time.

Administration. The highest authority at *Boimondau,* as in all the other *communities of work,* and for that matter in any cooperative community, resides in the General Assembly of all members. The Assembly meets twice a year. The members keep in constant touch with each other, and with the affairs of the *community,* through the Assembly of Contact. The Assembly elects a Head or "Chief" of Community who is in charge of all business and communal affairs. His term of office is three years. Among his duties and privileges is the right of veto of General Assembly decisions. In a case where the

Assembly decides not to yield to his veto, he may do either of
two things: give in, or resign. The Head of Community is
assisted by the General Council. That body consists of seven
members and the heads of departments. It is elected by the
General Assembly for one year. Within the General Council,
the Head of Community together with the section managers
and eight members, among them two of the wives, form a
Council of Direction.

Also elected by the General Assembly are the two supervisory bodies, the Court and the Commissioner of Control. The
Court consists of eight men charged with settling conflicts and
disputes between members, a task in which they are guided
mainly by the "common ethical minimum" and common sense.
Husband and wife always face the court together, even if only
one of them has become guilty of a violation. The Commission
of Control is a general supervisory body. Its main job is to
remove the snags from business and community affairs.

The decisions and elections of all administrative bodies must
be unanimous. If a decision cannot find the support of all,
it is postponed until unanimity can be achieved. Children,
apprentices, and *postulants* have no vote, though they are
allowed to participate in the deliberations.

The unit that lends to the *community of work* its most specific character is the Neighborhood Group. Members who
happen to live in the same part of the city belong to such a
group. They are required to meet at the home of one of the
members, and do so as often as they like to. Here, all the
problems of the *community,* the professional as well as the
social, are first thrashed out in a most informal and personal
manner. The Neighborhood Group is actually the elemental
unit of the *community* and, in a way, the living cell of the
communitarian organism. It constitutes the most ingenious of
all the contributions made by the *community of work* to the

solution of the problem of the urban cooperative community. It makes it possible for the city worker to remain in-and-of-the-city and to regain, at the same time, the feeling of personal identity and the sense of belonging. It secures for him the essential boons of community without exposing him to the pitfalls of withdrawal, of separatism and of "ghettoization."

Finances. When Mrs. Bishop first saw the people of Boimondau, she was struck by their "busy, carefree and free" manner, and she noticed their "self-assured and happy look." This impression she supports by quoting testimonials from the members themselves. For the more statistically minded, she quotes figures which confirm the material success of *Boimondau.* The capital at Barbu's disposal at the start in 1941 amounted to 300,000 francs. In 1943, the Germans destroyed the factory. The members succeeded in saving the machines, and in 1944 they began again from scratch. By 1946, their capital investment amounted to over eight million francs. They repaid to Barbu his original outlay plus guaranteed purchasing power, or a total of 565,000 francs. Further repayments, in replacement of machines, were made to Barbu, amounting, in 1949, to altogether 7,600,000 francs. The production of watch-cases, which in 1946 reached the figure of 221,296, rose to 339,667 in 1949, and the net sales returns (after taxes) from 28,000,000 francs in 1946 to 98,567,009 francs in 1949. These achievements were possible, Mrs. Bishop suggests, because *Boimondau* adopted, from the beginning, the most advanced methods of production, including scientific planning, assembly line technique and even time motion. Today, *Boimondau* accounts for twenty per cent of all the watch-cases manufactured in France and is one of seven—out of a total of twenty-seven—French watch-case factories that operate on a mass production basis. It is in every respect a "large" concern. Indicative of the financial solidity of

Boimondau is the fact that it already has helped four other communities to get started. It was also able to offer financial help to ten other communities which ran into difficulties.

The Communitarian Federation

Communities of work affiliate with a Federation, the *Entente Communautaire,* which has its headquarters in Paris. The Federation acts as a coordinating agency for all *communities of work.* It represents them in their dealing with government agencies; it assists them in drawing up their by-laws; advises them in financial matters; and it publishes the periodical *Communauté* as well as other communitarian literature.

There were altogether sixty communities in contact with the Federation in the Spring of 1950, the time of Mrs. Bishop's visit. At the last annual General Assembly of the Federation, in October 1950, Michel Anselme, the then general secretary, presented a report—published in the December 1950 issue of *Communauté*—which listed eighty-three communitarian projects, in different stages of development. Obviously, this means that the rate of increase in the number of communities is considerable. Of the eighty-three communities listed, only seventy indicated the size of their membership, which adds up to 1,460 households, *Boimondau,* the largest, accounting for 150. The three next in size, *D'Autryve* (wearing apparel), *Le Bélier* (watchcases) and the Swiss *Porteurs de Lait* (milk distributors), each had a membership of 100 households. Of the remaining, eleven showed a membership of thirty or more, and fifty-five of less than thirty families. Thirty-five of these had less than ten households, and the smallest groups counted as few as two families. Listed in order of frequency, the trades and occupations of the communities were as follows: mixed farming (13); construction (6); watch-cases (5); footwear, mechanics, woodwork and furniture (4 of each); vineyards, metal construction, cabinet work and apparel (2 of each); and

carpentry, woodwork, knapsacks, locksmiths, dolls, picture
frames, ironmongery, coppersmiths, milkmen, lathe-work, educational games and recitals (one of each). It seems as though
any trade and any occupation could serve as the economic
basis for a *community of work*.

According to their origin, the report distinguishes three
different types of communities: (1) those converted from private enterprise; (2) those initiated by a group of workers; and
(3) rural communities. Many of the enterprises still are in their
initial stage and the question arises of what criteria to apply
in order to assess their status and potentialities. M. Anselme,
in his report, considers the following factors to be crucial for
success: (1) possibility of attaining a size of at least thirty
households; (2) physical conditions favorable to expansion;
(3) availability of sufficient technical and business skill; (4)
certain personal qualities of the members. It is interesting to
note that of the nine communities which in addition to the
"Big Four" satisfy these criteria, all were converted from
private enterprises .Least promising, in the light of the above
criteria, are the rural communities. None of them has as yet
reached a size larger than ten households, but their number,
as the quoted statistics show, is relatively large.

Drawbacks

In tracing the main features of the *community of work* we
have followed Mrs. Bishop's pioneering narrative quite closely.
It is as vivid and readable as it is factual. There can be no
mistake about the deep sympathy with the aims of the communitarians that animates her report. Does not the elation that
comes from watching men carve out for themselves a new,
more decent and more satisfactory way of life, endow her observations with more enthusiasm than the facts warrant? The
careful reader will find, however, that Mrs. Bishop is by no
means unaware of the difficulties besetting the *communities of*

work. "In a young community," she states, "physical and material sacrifices are tremendous." Since, as we have seen, most of the *communities* are young as yet, the hardships must be assumed to be general. It is not surprising that relatively few are succeeding. It is not easy to find in the same man high technical skill and a genuine communitarian spirit. Yet, this is precisely the kind of man on whom depends the success of a *community of work.* If it tries to make light of either of these two basic attributes for the sake of the other, it may fail either as an economic enterprise or as a community.

The most crucial kind of internal trouble arises, however, from some basic attitudes of the women. It may be not altogether correct, as Mrs. Bishop maintains, that the "average European wife wishes but one thing: that 'her man' be happy;" but it may well be true for the average French woman. If she feels that building a *community of work* will make him happy, she will readily put up with hardships. She may feel neglected when he gives all his time to the community, and she may go so far even as to sabotage the formation of a Neighbor Group for fear that, "with so much intimacy," her husband may get himself entangled with another woman. Yet, on the whole, she does her share. It is only when there are children that she may balk. She will resent nothing more than seeing them neglected by their father, and she may confront him with a choice between his home and the *community.* Conflicts of this kind, as we know from the early history of the *Kvutza,* are apt to destroy many a struggling group.

The self-criticism of the leadership of the communitarian development is even more comprehensive. In a very outspoken editorial in the June-July 1950 issue of Communauté, entitled "Raison d'être . . .," M. R. Roquette, associate general secretary of the Federation, takes account of certain basic criticisms of the movement by disillusioned sympathizers. Contrasting the intensity of efforts with what to them seems like a paucity of achievement, they conclude that the "communitarian formula" has failed. They point out that only a few of the groups organized by the workers themselves are doing well; that many merely vegetate without any promise for the future; that some have disappeared altogether; and that those enterprises that had done well before conversion into *communities* now encounter considerable financial difficulties. In short, all that remains from a movement that set out to establish a new order is a few isolated instances of fairly prosperous *communities*; and even their future appears to be by no means assured. M. Roquette does not reject these criticisms as unjustified. He merely reminds the critics that the fault is with their own over-optimism rather than the movement. It should not be forgotten, he stresses, that the whole development did not get really started until five years go. For that relatively short period of time the success, although below the high expectations of some enthusiasts, need not be considered negligible. It is disappointing only to those who never stop looking for panaceas and fool-proof formulas. The *community of work* is not and does not offer any such formula. It is a movement that gropes after a new and better coordination of industrial work and social existence. It is a sober search, and the people engaged in it sometimes learn more from their failures than from their successes.

A Significant Advance

The gist of M. Roquette's argument, as paraphrased above, may seem to tone down somewhat the impression conveyed by Mrs. Bishop's book. It may be that the movement is not forging ahead as indomitably as some parts of *All Things Common* would want to have it. Still and all, there can be no doubt about the fact that the *community of work* carries on a sociological experiment of significance. Dissatisfied with their two-fold alienation, from the means of production and from the normal chances of personal fulfilment, they attempt a fundamental change of their situation. They proceed by pooling their resources, and by turning work into an instrument of meaningful human existence. In the process, the members of the group are ready to experiment each with himself as well as with the possibilities of the new situation. They "control" each step of their experiment by checking it against the satisfactions achieved; and they "validate" it in terms of the degree of agreement achieved among themselves as to its merits. If the agreement is unanimous, they accept and institutionalize it; if not, they abandon it, and try again.

The nature of the *community of work* being that of a sociological experiment, it should be possible to evaluate it critically and compare it, with some precision, with related attempts. Such comparative evaluation would have to concern itself, as we suggest elsewhere,[6] with the felicity and fertility of the original premise of the experiment, as well as with the implicit or explicit methods of control and validation. In the case of the *community of work,* this would require a more incisive, and a more systematic study than Mrs. Bishop's report intended to be. On the basis of the available facts we might compare critically the premise of the *community of work* with those of the two main related undertakings, the Industrial Cooperative on the one hand, and the *Kvutza-type* of modern cooperative community, on the other.

From what has been said, it should be clear that the premise of the *community of work* is definitely more felicitous than that of the Industrial Cooperative. The latter proceeds from the assumption that work relations can be effectively organized on a comprehensive cooperative scale without regard for social relations. The relatively poor showing of the Industrial Cooperative after a century of effort indicates the weakness of this assumption. No such discrepancy between work and social

[6] See "Utopia and Experiment" in this volume.

relations characterizes the basic approach of the *community of work.*

The *community of work* arises on grounds more fertile also than those of the *Kvutza.* The specific premises from which the latter proceeded suggested itself by the given circumstances. Since individual farming moved too slowly and proved so wasteful and difficult as to appear all but prohibitive, there was no alternative to the attempt at group farming. The community that grew out of the ensuing experiment was thus confined to agriculture. The fact that industry was added later on did not actually modify the essentially rural character of the *Kvutza.* As shown by the experiment of the only City *Kibbutz, Ef'al,* the *Kvutza*-type of community does not appear to be transferable to urban conditions. The *community of work,* on the other hand, grew out of an urban setting. It proceeds from the premise that by moulding his work as well as his life along communal lines the industrial worker can achieve a sense of belonging and of personal worth without necessarily withdrawing from the city. As the relatively large number of mixed-farming *communities of work* shows, this premise is not without possibilities of application to rural conditions. In this sense, the *community of work* surpasses, at least potentially, the *Kvutza* in range of applicability. Thus it points toward a significant advance in the development of the modern co-operative community.

PART II

10

THE SOCIOMETRIC TEST: ITS MERITS
AND ITS LIMITATIONS

Sociometry and the "Open" Community

The use of sociometric techniques is of a relatively recent date. Advancing, both methodologically as well as technically, into areas previously little explored, sociometrists have been careful not to venture too far afield. They have confined their investigations mainly to small and relatively simple units, such as the so-called closed groups. Only recently has it been felt that sociometry had matured, and was capable of taking on more complicated tasks. This feeling was expressed forcefully by J. L. Moreno. In the conclusion of a survey of the origins of sociometry, he stated: "The time has come, after twenty-five years of research in catacombs, such as prisons, hospitals, reformatories, schools, that sociometry moves from the closed into the open community."[1] There can hardly be any doubt that this advance is highly desirable. What might be questioned is whether it will be possible to step, as Dr. Moreno recommends, from the "catacombs" directly "into the midst of every town, every region, county and state." These, it would seem, are areas of a magnitude still beyond the capacity of sociometric techniques. A further refinement of existing procedures and additional devices are apparently needed if sociometrists want to assure such advance. At this stage, it should be of great advantage already to study groups relatively small in size but "open" in the sense of any "normal" human society.

---

[1] J. L. Moreno, "Origins and Foundations of Interpersonal Theory, Sociometry and Microsociology," *Sociometry*, XII, No. 1-3 (August 1949), p. 254.

Groups of this kind are the modern cooperative communities which exist today in many parts of the world.[2] Field Studies of some of these communities in the U.S.A., Canada, Israel, and France have been conducted by the Group Farming Research Institute, under the direction of this author. The use of the sociometric test in these investigations helped reveal certain of its limitations as well as virtues that might not have been apparent before. Such experience has led to refinements and extensions in the use of sociometric techniques that should be of interest to sociometrists intent on moving into the "open" community. It is the purpose of this essay to record them briefly.

The Sociometric Test

This test is designed to reveal the actual personal interrelations existing in a given group of people. If correctly used, it takes the clue to its application from definite needs of the people, and elicits responses which serve to disclose discrepancies between the manifest group structure and the underlying network of interpersonal attractions and repulsions. Often these discrepancies prove to be the main barriers between people's needs and their satisfactions. If this is the case, the results of the test are used to improve the situation and to make it yield the desired satisfactions.

The sociometric test is genuinely sociological, for it conforms with the method of sociological experiment (as described in "Utopia and Experiment" in this volume) and thus "accords" with the "properties" of social phenomena. To be sociological, any test, it would seem, would have to do so. Contrary to some implicit, and even explicit claims made by Dr. Moreno, the sociometric test, however, is not in itself

[2] Cf. H. F. Infield, *Cooperative Communities at Work* (New York: Dryden Press, 1945 and London: Kegan Paul, 1947). See also: "Some Recent Developments in Cooperative Farming" in this volume.

a sociological experiment.[3] It is rather a product of an imaginary experiment. The experiment stipulates that to be truly sociological, it must leave unimpaired man's fundamental characteristic, his capacity for deliberate action. As it is designed to include all mankind, it is of a scale on which only a superhuman being, or God, could operate.[4] The assumption which this imaginary model of an experiment is to test might be stated as follows: For the most part, relationships between human beings are such that some basic needs fail to be satisfield. To effect a change for the better, it is necessary to restructure the whole of existing interpersonal networks. It thus becomes imperative, first, to examine into the nature of present relationships, and second, to bring about desired changes in accordance with the choice preferences expressed by the people. In this connection, the construction of an instrument capable of registering the choices of mutual attraction and rejection becomes of crucial importance.

In the construct of his imagined universal sociometric test, such considerations may have moved Dr. Moreno to choose for the first comprehensive publication of his sociometric studies a title that since has puzzled many a serious student.[5] To Dr. Moreno, however, it seemed an appropriate title, for it helped dramatize the findings of his model experiment. "Who Shall Survive?'—, such a question was properly asked in a society satisfied with wasting a very considerable part of its human resources. By contrast, it would be meaningless in a

---

[3] Cf. J. L. Moreno, "Sociometry and Marxism," *Sociometry, loc. cit.,* p. 114.

[4] See Moreno's statement: "Metaphorically, God might be called a sociometrist on a cosmic scale," "Sociometry and Marxism," *loc. cit.,* p. 118.

[5] J. L. Moreno, *Who Shall Survive? A New Approach to the Problem of Human Interrelations.* (Washington, D.C.: Nervous and Mental Disease Publishing Co., 1934.) Also L. von Wiese, "Soziometrik," *Koelner Zeitschrift fuer Soziologie,* I, 1. A translation of this article under the title "Sociometry" appeared in *Sociometry, loc. cit.*

"sociometric" society where no one would be cast out, and all would be given an opportunity to participate to the best of their abilities—or to "survive."

Apart from its theological fortuity there is nothing inherently unscientific about a mental model of this sort, even though it is obviously "unreal." Max Weber's "ideal type," [6] too, is a construct, as are the mental models of multi-dimensional universes used for heuristic purposes by mathematical physicists. A mental construct can be useful even though it is not found in "reality." But applied to concrete situations, the sociometric test "works." It is a device of sociological examination, though not an experiment itself.

Cooperative Communities are Instances of Genuine Sociological Experiment

Yet there are in empirical reality experiments which conform with the sociological canons: the modern cooperative communities, particularly their most comprehensive realization, the Israeli *Kvutzot,* are such experiments. [7]

They are so in the sense, and to the degree, that they are spontaneously initiated by people finding themselves unable to satisfy some of their basic needs within the dominant social structure. By deliberate action they drastically modify the basis of their economic relations. They pool their resources and decide that henceforth they will own everything they possess in common. This fundamental modification necessitates changes in their social interrelationships. Step by step such changes are introduced with the consent, and under the control,

[6] "In its conceptual purity, this mental construct (*Gedankenbild*) cannot be found empirically anywhere in reality. It is a utopia." Max Weber, *The Methodology of the Social Sciences.* Translated and edited by Edward A. Shils and Henry A. Finch. (Glencoe, Ill.: The Free Press, 1949), p. 90.

[7] Cf. H. F. Infield, *Cooperative Living in Palestine* (London: Kegan Paul, 1946, and New York: Henry Koosis & Co., 1948).

of the members of the group, and are institutionalized as soon as they are found validated in terms of satisfactions yielded. Depending on circumstances and personalities, such validation may lead to extending cooperation only partially beyond production; in other cases, as in the *Kvutzot,* cooperation may become all-inclusive and decisively shape all aspects of social organization.

The actual conduct of these experiments offers valuable opportunities for sociological research. It has already produced considerable insight into issues bearing on some central problems of our present-day society. To extract this insight in a manner compatible with scientific objectivity, adequate observational tools are necessary. These tools must satisfy two basic requirements: they must be commensurate with the nature of the phenomena in question; and they must be as precise as possible. Since, sociologically speaking, cooperative communities are functional groups, the facts that any research will endeavor to ascertain first are those related to group structure, the status of members, the degree of group integration, and so on. It is clear that the sociometric test is so essential for this purpose that it would have to be invented if it were not available. From the beginning, this author has therefor made extensive, though not exclusive, use of it in his field studies.[8]

The Merits of the Sociometric Test

(1) The general technical simplicity of the sociometric test is certainly one of its most attractive features. The equipment consists of paper and pens or pencils. Technical simplicity of this kind is of great importance in all social science research which, probably for some time to come, will have to carry on with limited funds. It is of particular advantage, however, in cases of groups existing under the most primitive conditions.

[8] Cf. "Current Research," *Cooperative Living,* I, 1.

Another advantage of the test is the relative ease with which results can be tabulated.

(2) The issue of membership selection is probably the most fateful one in the stage of cooperative group formation.[9] The use of the sociometric test made it possible to clarify this issue in the case of Macedonia, the cooperative community formed originally by Morris Mitchell with a group of local farmers near Clarkesville, Georgia. The original group was replaced by one of persons from all parts of the U.S. whose common bound was pacifism, the belief that cooperative community living can serve as a "moral equivalent to war." [10] Putting devotion to their faith above any other consideration, the group found it unethical to deny admission to anyone professing pacifism. Consequently, people kept drifting in and out, and activities, hopefully initiated, had to be abandoned, often with considerable losses. Were it not for a relatively solid small nucleus the group probably would have met the fate of all communities negligent of the need for selection and would have become one more abortive attempt at cooperative group formation.

In spite of hardships and extreme difficulties of a personal nature the illusion prevailed, however, that the group was growing. Here the sociometric test proved more helpful than even the best verbal argument would have been. The results of the test showed convincingly that what existed at Macedonia was not even a group in the proper sense, to say nothing of a cooperative community.[1] Stirred by these findings the leadership took steps to put the organization of the group on a more

[9] Cf. Henrik F. Infield and Ernest Dichter, "Who is Fit for Cooperative Farming?" *Applied Anthropology* (January-March 1943).

[10] See Art Weiser, "The Macedonia Community." In: Henrik F. Infield and Joseph Maier (eds.), *Cooperative Group Living, An International Symposium on Group Farming and the Sociology of Cooperation* (New York: Henry Koosis & Co., 1950). Also "Quantitative Group Comparison," in this volume.

[11] Cf. David Newton, "Problems of a Cooperative Colony," Arthur

solid basis. First among these steps was the requirement of a probation period for all new applicants. The group is still relatively small, and there are still hardships, but today there are also signs of permanence.

(3) The possibility of utilizing the sociometric test for administrative action based on objective data instead of chance was revealed in the case of the veterans' cooperative land settlement, Matador, in Saskatchewan, Canada. Matador[12] was the first of several such settlements formed with the active support of the provincial government. The Cooperative Commonwealth Federation (CCF) party which came to power in Saskatchewan in 1944, saw in cooperative farming a possible solution to the problems of the unmarried veterans who, wanting to settle on the land, found themselves for different reasons incapable of acquiring a family farm. The CCF therefore encouraged the veterans to pool their resources and to start farming together. The use of the Veterans' Land Act money ($2,320. per veteran) for this kind of settlement met with objection on the part of the Federal Government of Canada. The CCF nevertheless went ahead with its plans. A first group, consisting of seventeen veterans, was moved to Matador, on provincial Crown Land, and began farming operations.

Meanwhile, negotiations about the use of the veterans' money for cooperative settlement went on between Saskatchewan and the Canadian Government. Pressed by public opinion, the federal authorities at one point were ready to grant permission, but only on the condition that the size of any such settlement be limited to not more than fifteen veterans. Ready

Naftalin a.o. (eds.), *An Introduction to Social Science* (Chicago, 1953), III, 321.

[12] See "Saskatchewan" in this volume, also "Sociometric Structure of a Veterans' Cooperative Land Settlement," Sociometry Monographs No. 15 (Beacon House), 1947; and Harold E. Chapman, "Report on Recent Developments in Saskatchewan," *Cooperative Living*, I, 2 (Fall 1950).

to compromise, the CCF faced the problem of picking the two men who would have to go. In view of the publicity to which the whole undertaking was exposed, the decision was a weighty one. Adverse criticism could be avoided, or at least made unreasonable, only if the decisions were based on objective criteria. The sociometric test yielded such criteria. The CCF was thus able to avoid making a haphazard and risky choice but instead could determine, in terms of the raw status and mutual scores, the two men most marginal in the group whose leaving would least damage it. From the point of view of the group itself, the question it was forced to face was: who shall not survive. Here, too, the use of the sociometric test proved to be helpful. It made it possible to keep the painful consideration on an impersonal objective level.[13]

(4) The use of the sociometric test in the quantitative comparison of the two communities, Matador and Macedonia, maybe considered a contribution to social science methodology.[14] The comparative study of groups, societies, cultures is one of the main objectives of sociology and even more so of anthropology in their efforts to arrive at a body of general principles. To comply with the requirements of scientific accuracy the comparative method must demonstrate its capacity for producing objective criteria and verifiable propositions. It is the common complaint of the most advanced workers in the field, however, that little headway has been made in this regard.[15] The chief obstacle it seems, is the lack of a standar-

---

[13] After weighing the decision for some time, the CCF took a different way out. It assured payment of the grants out of its own funds should the negotiations with Ottawa fail.

[14] See "Quantitative Group Comparison" in this volume.

[15] For the sociological side see Samuel S. Stouffer, "Some Observations on Study Design," *American Journal of Sociology (January* 1950) For anthropology, see Clyde Kluckhohn, "The Personal Document in Anthropological Science." In: *The Use of Personal Documents in History, Anthropology, and Sociology* (New York: Social Science Research Council Bulletin, No. 53, 1945).

dized unit of reference. Without it comparison can be neither significant nor consistent.

The application of the sociometric test to the two communities indicated the possibility of a methodological advance. The basic unit of reference, interpersonal choice, was the same in both cases. The comparative use of the indices this unit produced had been facilitated by the limited but adequate statistical standardization offered by Bronfenbrenner.[16] The technical pre-condition for an attempt at quantitative group comparison thus appeared to be given. A circumstance stimulating the attempt was the fact that, although the two groups differed slightly in size of membership, exactly the same number, twelve, happened to take part in the tests. Both groups being cooperative communities, their structure was of the same socio-economic kind. They differed in motive of their origin, background of members, and professional preference for cooperative living as a goal in itself.

The results of the study are presented in the next essay. A few of the most general data are cited here for the purpose of illustration. Inspection of the mutual choice sociograms, based on the three criteria of "work," "leisure time activities," and what for the sake of brevity may be called "mutual confidence," yielded the following scores:

	Work	Leisure	Mutual Confidence
Matador	10	7	5
Macedonia	7	9	8

These scores were taken to indicate that while Matador was almost optimally teamed for work, was only fairly well so on leisure time activities, and was close to poor on mutual

[16] See Urie Bronfenbrenner, *A Constant Frame of Reference for Sociometric Research*, Sociometry Monographs, No. 6.

confidence, Macedonia on the other hand appeared to be strongest on leisure time activities, somewhat weaker on mutual confidence, and weakest on work. Based on these scores, it was possible to make tentative estimates of the comparative chances for survival of the two groups, both of which at the time still were in their early formative stages. It was possible to predict that if the living standards, due to material hardships, would sink below a certain minimal level, in the case of Matador the personal ties would prove too weak to hold the group together. Macedonia, on the other hand, would have a much better chance for survival even under conditions which would appear to the Matador group as unbearable. The findings, furthermore, helped indicate the lines of action to be taken in order to improve either of the two groups chances for survival. They showed that what one group needed was strengthening of personal ties—"to know each other better," as one of the Matador group put it—and that what the other had to do was to improve its work organization.

(5) Finally, the sociometric test, when applied to one of the Training Farms mentioned, proved helpful in elucidating certain educational issues. These Training Farms, the number of which has been drastically reduced since, are essentially part of the preparatory system of the *Kvutzot*. They are occupied by young Zionists prior to their emigration to Israel. Organized along lines virtually identical with those of a *Kvutza* they differ from them mainly in that admission is less strict, and that membership is limited to the time of training, normally one year. The task of the farms is to prepare the pioneers for both cooperative work and cooperative living. Since all of the graduates plan to settle in a *Kvutza,* proficiency in both is equally important.[17]

At the farm in question, marked uneasiness had been caused in the group by an unusually large number of resigna-

[17] Cf. "Effective Education" and "Testing of a Pioneer Training Farm" in this volume.

tions. The results of the sociometric test confirmed that the state of the group was poor indeed. The raw status scores of the 32 members showed the following distribution: "Star": 2; "Upper" Bracket: 5; "Middle Bracket": 8; and all the rest, 17, or over 50 per cent "neglectee." There was certainly good reason for uneasiness.

Consultation of the individual scores of the two "stars" revealed that of the great number of choices they had received, the largest part by far had been registered on one criterion, work. The one "star," A, with a raw status score of 25, showed 18 choices on work, 4 on leisure activities, and 3 on mutual confidence. Similarly, of the 23 choices received by the second star, B, 17 were on work, 4 on the second, and only 2 on the third criterion. In other words, both were "stars" or close to "stardom" only as far as work was concerned; but they were "neglectees" on all other criteria. This gave an unmistakable clue to what was probably the main reason for the poor state of affairs. The leadership that had developed in the group evidently was based almost exclusively on work relations, at the expense of all personal mutuality.

The presentation of these scores to the General Assembly produced an eminently" cathartic" effect upon the members. In a session that followed they decided to take immediate steps toward remedying the situation.

Limitations and Shortcomings

(1) As against these good uses to which the sociometric test may be put, its application to the cooperative groups listed above revealed also several of its limitations and shortcomings. The first is of a general technical nature. The main part of the sociometric tests consists of a questionnaire, and thus depends on articulate verbalization. The test, consequently, shares with other devices of standardized observations based on verbal responses all the difficulties implied in semantic clarity, veracity of responses, and validity of interpretation.

In the case of cooperative communities, some of which, as for instance in Mexico, count illiterates among their members, these difficulties would become particularly acute. They would make the application of the test, at least in its present form, virtually impossible. Cases of resistance to the test because of verbal difficulties were encountered in several of the groups. To overcome such resistence, resort had to be taken in one case to individual application of the test and to oral translation of the questionnaire into the idiom with which the individual was most conversant.

Another technical shortcoming lies in the nature of the questions asked. A certain reluctance against stating preferences and rejections in terms of personalities can be noticed in almost every group. There is fear of disclosure and of ensuing enmities. This fear must be particularly strong in any group in which participation is so all-inclusive as in a cooperative community. The question of rejection has hardly any chance at all of being answered under these conditions; and even the question of first, second, and further choices is, since anonymity would contradict the sense of the test, often answered, only after assurance of confidenc. None of this adds to the veracity of the responses.

(2) A more specific technical difficulty lies in the degree to which the application of the test depends on the skill of the investigator. This is so because the phase preceding the test, the "warming-up-"process, is at least as important as the questionnaire itself. To initiate this process seems to be relatively easy in groups which by virtue of their situation find themselves deprived of a considerable part of their capacity for deliberate action. In closed groups of this kind, in the sanatorium, the reformatory, and to some degree in the school, the control of the situation lies in the hands of some person of authority, the physician, the housemother, the teacher, who usually meets with no difficulty in applying the test.

No controlling authority of this kind exists in cooperative

communities. By principle and practice, they exclude authoritarian manipulation of the members. To secure any response at all the investigator must be capable of winning the confidence of the group, and of stimulating it into accepting the test as a matter of group action. This calls for skills and talents which need training. At best, only rudiments of such training are offered in the courses of colleges and universities. Its lack makes the outcome of the test highly dependent on the presence of mind and inventiveness of the investigator. This presents a serious obstacle to large-scale application of the test and explains the reluctance of sociometrists to venture into the open community.

(3) A further, mainly methodological limitation can be traced to the fact that the sociometric test until now has been applied most widely to closed groups. Significant informal networks of interpersonal relations usually develop in this kind of group around a limited number of criteria. When Bronfenbrenner undertook his very helpful standardization, he did not extend it beyond three criteria. That was all that was needed to satisfy "current sociometric practice." [18] In the GFRI's research it was not possible to keep within the limits of this practice. For cooperative communities, the adoption of "job-analysis" required a minimum of eight criteria. This created a special problem. To restrict the questionnaire to fewer than eight criteria made it incomplete; to extend it meant foregoing the advantage offered by Bronfenbrenner's standardization. In the case of the quatitative comparison of Matador and Macedonia the problem was solved by "matching" the two groups on three criteria.[19] However, this could not be considered to be more than a temporary expedient. Further methodological advance will be necessary, to make standardization fit the requirements of "open community" investigation.

(4) What may be called "status anxiety" was found to be

---

[18] See Urie Bronfenbrenner, *loc. cit.*

[19] For details, see "Quantitative Group Comparison" in this volume.

one of the main sources of a particularly strong resistance against the sociometric test. Resistance of this kind was encountered in the *Kvutza Maagan,* in Israel. All the other communities mentioned before were still in the formative stage at the time the test was administered. *Kvutza Maagan,* however, had been in existence for about ten years at the time of investigation. Solidification had progressed to the point where the "careers" of members had fallen into well defined grooves. In a cooperative community, such solidification produces, it seems, a certain amount of status anxiety, which is rarely admitted openly, or even covertly.[20] In view of the formal equality practiced by the *Kvutzot* anxieties of this sort need not come to the fore at all. They may be aroused, however, when, as in the application of the sociometric test, wishful deception about one's own status is forced to face reality. Many persons apparently prefer to keep the facts veiled rather than admit that their frustrations have a basis in fact. A "new start" being extremely difficult at this point of development, conscious admissions may leave open only one course of action, resignation from the group. It was for this reason, we believe, that only a small number of members, who otherwise willingly responded to other observational devices, cared to answer the sociometric questionnaire.

(5) Application of the sociometric test to cooperative communities revealed still another shortcoming, a certain theoretical one-sidedness. As we have seen, sociometry proceeds from the assumption that one of the main barriers between the people's needs and their satisfaction lies in the discrepancy between the manifest group structure and the underlying networks of interpersonal relations. Revelation of these informal

[20] Cf. "Pains of Growth in Israel" in this volume.

[21] This may call to mind the therapeutic mechanism of psychoanalysis: the cause of neurosis, the trauma, is made non-malignant by conscious recall. It might be that the formulation of the basic assumption of sociometry has been influenced by the instance of psychoanalysis. Actually, however, the mechanism is here reversed: it is the

networks by means of the test is supposed to remove the barriers, and to make possible satisfaction of the needs.[21] This ameliorative effect is one of the good uses to which the test has been put in the case of many of the closed groups, and even, as we have seen, in the cooperative community, though only in the formative stage. In a solidly established cooperative community, however, the relation between formal and informal group structure appears to be reversed. Modern cooperative communities, as they exist today, represent a deliberate experimental departure from society as generally known. In this sense, solidification means that the experiment succeeded, and that the new pattern of life was found to be satisfactory. What we face here is a formal organization which facilitates the satisfaction of the common needs, and informal networks such as cliques, factions, partisanships which tend to obstruct such satisfaction. To "sociometrize" a group of this kind might easily lead to dire consequences. Experience shows that whenever these informal networks manifest themselves in a cooperative community they tend to disrupt it. In other words, application of the test to a cooperative community does not appear to verify the basic theoretical assumption of sociometry. It rather suggests the narrowness and the onesidedness of the assumption and its re-statement in broader, conditional terms. The sociometric test, it might be said, is designed to reveal the networks of interpersonal relations underlying the formal group structure. In themselves these networks are neither good nor bad, and neither better nor worse than the formal structure as regards satisfaction of the common needs. Whether they are the one or the other depends in each case on the nature of the needs, the situation, and the formal structure.

To illustrate: if the formal structure, as for instance in the

manifest organization which is malignant, and the latent networks which are supposed to bring the cure if made manifest, or institutionalized.

reform school, is coercive and does not allow enough range for personal choices, institutionalization of the informal networks will have a liberating effect. If, on the other hand, the formal structure, as for instance in a cooperative community—or for that matter in any group which like a scientific body has for its purpose the pursuit of a common goal—is framed so as to make possible the widest range of personal choices, the informal networks will have the opposite, impeding effect. In this case the results of the test will not be used to institutionalize, but rather eliminate, or at least to control the informal networks.

(6) A final, but most significant, limitation is related to the scope of the sociometric test. The test, as we have seen, yields indices of group structure, status, and degree of integration. That is all the test is designed to do, and it would be unjust to blame it for not doing more. Social structure and related phenomena, though essential for the comprehension of any group, nevertheless explain only one basic aspect of its funtioning. To observe, and if possible to measure, other important factors, such as manner of origin, backgrounds, needs of members and so on, other devices are needed in addition to the sociometric test. Moreno himself apparently recognized the need for other procedures, by stating that "Sociometrists, in order to attain the fullest usefulness of their instruments should combine sociometric tests on the choice and on the reality level with psycho-socio- and axiodramatic procedures and should always be ready to make modifications in favor of the community of people to which they are applied."[22]

The need for additional devices, not necessarily of the type recommended by Moreno, became evident when the scores of the sociometric test revealed significant differences between two of the training farms referred to above. Both groups had been in training for about the same length of time; both had the same kind of membership; and both were run in the same

[22] See J. L. Moreno, "Origins and Foundations," *loc. cit.,* p. 254.

way. In one case, however, only twenty percent rated as "neg-lectees"; in the other, as many as fifty percent of the member-ship showed up in the same bracket. If these percentages were taken as indices of group integration, one showed a fair de-gree of integration, while the other was virtually falling apart. The sociometric test could well help ascertain this startling difference, but it could not be used to reveal the reason for it. To attack this aspect of the problem other procedures had to be devised, such as the "biographical group interview," the "personal questionnaire," the "cooperative potential test," and the "obstacles test." The fullest use of these latter devices was made in the study of the French *Communities of Work.*[23]

[23] Cf. "Experimental Groups and Sociological Counselling" in this volume.

11

QUANTITATIVE GROUP COMPARISON

Matador and Macedonia

Group comparison based on objective, measurable data is probably one of the more significant fields of investigation which needs to be cultivated if sociological research is to become cumulative. Opportunity for such comparison appeared to be given when the results of two studies made by the Group Farming Research Institute were examined. The first of these studies was that of a veterans' land settlement at Matador, in the Canadian province of Saskatchewan; the second, of a pacifist group settled at Macedonia, in the state of Georgia.

These two groups differed in respect to origin, background, and attitude. The one, Matador, was Government sponsored, and its members were all men; the other, Macedonia, owed its origin to private initiative, and consisted almost wholly of married couples. Only one of the men at Matador had gone to college, but they were all born on farms and most of them were experienced farmers; most of the members of Macedonia had grown up in towns or cities, farming was new to them, but virtually all of them, women as well as men, held college degrees. And of course, the veterans of Matador had conformed willingly with the accepted mores, and had volunteered for service in the armed forces; while the pacifists of Macedonia had gone to the CPS camps and even to jail rather than to comply with the rules and regulations of the Selective Service Act.

But there were two essential aspects in which the two groups were similar, if not identical. These were their socio-economic structure, and the size of membership. Both were cooperative communities of the modern type which derives its tenets from the Rochdale principles; and both practiced, at the time they were visited, about the same degree of comprehensive cooperation. The number of members was in each group exactly the same—fifteen.

These essential similarities suggested comparison; but it was the use of the same techniques of group investigation which made it feasible. The techniques were those of the biographical group interview, and the sociometric test.

Matador, in Saskatchewan, Canada.

A description of the two communities may be helpful in making the results of the two studies and their comparison intelligible. Matador, which has been dealt with elsewhere,[1] may be treated briefly. Macedonia will be described in greater detail.

The establishment of the Matador Veterans' Cooperative Land Settlement is closely connected with the coming to power of the Cooperative Commonwealth Federation (CCF) in Saskatchewan. In the elections of 1944 this party won forty-seven out of fifty-two seats in the legislature of the province. At a conference, called in August of the same year, the Minister of Reconstruction, the Hon. John H. Sturdy, proposed to offer returning veterans an opportunity to settle cooperatively.

The offer appealed strongly to single veterans who had a harder time in finding individual farms. Twenty-six of the men who had responded to the invitation of the Government were brought together in Regina, in April of 1946. In a week long session the men were given opportunity to discuss all

---

[1] See "Saskatchewan" in this volume and Henrik F. Infield, "A Veterans' Cooperative Land Settlement and its Sociometric Structure," *Sociometry*, X (February, 1947) or *Sociometry Monographs*, No. 15.

problems related to the new settlement. Seven of the attend-
ants dropped out; the remaining nineteen were out at Matador three weeks later. Soon, four of the men left.

The conditions the men faced upon their arrival could
hardly have been more primitive. There were no buildings,
and no means of communication. The nearest dirt road was
five miles away, the closest neighbor lived at a distance of
six miles. For the first few weeks, their only accomodation was
a caboose.

At the time the study was made, July, 1946, two temporary dwellings had gone up. One served as a common dining
hall, meeting place, and living quarters for a member and his
wife, who was in charge of the kitchen, the other as a dormitory for the men. The place still looked quite bare but activities were in full swing. Three crews had been organized: one
in charge of dismantling and moving buildings from a nearby
airport, a second for construction work, and a third, for farming operations. Several hundred of acres of land had been
broken by that time, with tractors leased from the Department
of Reconstruction, and with plows bought from farmers in
the neighborhood.

Although the plans foresaw a combination of cooperative
production with individual consumption—somewhat along
the lines of the Russian *kolkhoz,* or the Palestinian small-
holders' settlement—the group was forced to practice cooperation of an all-inclusive kind. With only two huts available
for fifteen men and one woman, there had to be a central kitchen, a common dining room, and a dormitory for the men.
Cooperation thus extended not only to work, but to all other
activities of the community, including recreation.

Macedonia, in Georgia, U.S.A.

The Macedonia cooperative community is located near
Clarkesville, Georgia, on a tract of about 1,000 acres of tim-

berland. The project[2] was initiated by Morris Mitchell, then Professor of Education at New College of Columbia University. The original motive in starting the enterprise was Dr. Mitchell's intention to show that private initiative could equal, if not surpass, the achievements of Government sponsored rural rehabilitation.

No more challenging place could have been found for such demonstration. Historically this area, situated at the foot of the Blue Ridge, had been a refuge for nonslave holding families from lower Georgia. These families had mined the originally rich soil, and, with seventy inches of rain annually, and a long growing season, they had lived in comfort. But they had also succeeded, within less than a hundred years, "to waste the natural resources to the point of almost complete depletion."[3] As degraded as the land, one might say, was the living standard of its people. The soil was obviously in need of reclamation, and the people of rehabilitation. Dr. Mitchell took on both jobs, each of them an heroic task in itself, simultaneously.

Dr. Mitchell began by buying 820 acres of land, and by inducing local people, one by one, to settle on it. At the height of the development, five families, three of them closely related to each other—altogether twenty-seven people—had moved to the place. Dr. Mitchell, who had invested close to $18,000 in the settlement and who kept investing part of his teacher's salary to keep it going, did not reside at Macedonia. A manager, appointed from among the members, was put in charge; but no decision was made without general consent. The men ran a saw-mill, started a dairy and an apiary, and opened a cooperative consumer's store. Housing and living improved

[2] Cf. Joseph W. Eaton, *Exploring Tomorrow's Agriculture,* (New York: Harper and Brothers, 1943), pp. 204-206.

[3] Morris R. Mitchell, "Habersham County in the Awakening South," *Progressive Education,* December, 1940.

considerably, and the intended demonstration seemed just about to succeed.

Then came the war, and slowly, for one reason or another, most of the members drifted away. At the time of the visit, February, 1947, only one of the families remained; but none of its members took part in the activities of the new group which had gathered at Macedonia.

It is with this new group that our study is concerned. The members of this group were quite different from the local settlers. Some of them had become acquainted with Macedonia as members of the Friends Work Camps which had helped to build the dam for a projected lake. Others had learned about the community in the Civilian Public Service (CPS) camps.

The manner in which the group met at Macedonia was quite informal. After the dispersal of the original community, Dr. Mitchell began to look for other settlers. Having himself become a pacifist, it was among adherents of this movement that he sought and found most eager response. This has to be understood in the light of the position in which pacifism in the Allied countries found itself during the last war. To prove truly conscientious, objection to war against Nazism required compensation by some sort of positive action for peace. It was the spirit of true community which produced harmony among men. The cooperative community—the most perfect expression of this spirit—was therefore the field of action which promised the most effective affirmation of the "moral equivalent to war."[4]

---

[4] In England, the more general aspects of community were stressed. Cf. "Comradeship of Peace," *The Community Broadsheet,* London, Spring-Summer, 1945. Also: *Comradeship Continued. A Practical Guide for Men and Women in a Post-War World,* Pamphlet, published by the Community Advisory Group, n.d.

In the U.S.A. and in Canada it was specifically the cooperative community which was favored.—Cf. the several issues of *The Communiteer,* The Newsletter of the Rural Cooperative Community Con-

The idea found organizational expression in the Rural Co-operative Community Conference in the United States and in the Community Service Committee in England. Both organizations kept in close touch with each other, and with the Canadian Fellowship for Cooperative Communities which, although not pacifist in principle, shared with them the interest in the cooperative community.

The publications of all three organizations reveal that serious consideration was given to matters of procedure. But the action that followed was disappointing, even to the faithful. The reasons for failure may be found, at least partly, in the following self-appraisal:

> The neurotics, the feckless and the failures may well become the special charge of community. But the community must be sure of itself before it can be of much use to these. Otherwise it will be swamped by the abnormal. That indeed has been the fate of some.
>
> Inevitably also the work-shy and those seeking a shelter from the wind have often made for the open door of the community—for the duration only. Nor have they contributed much to their keep, for community is not built by part-timers and itinerant ideologists.[5]

Some of the CO's had known of the cooperative community before entering theCPS camps. At the School of Cooperative Living, in CPS 30, Walhalla, Michigan, and in the Trenton, community group, at CPS 94, Trenton, North Dakota, serious study and discussion, and to some degree training in agricultural skills, were taken up in preparation for eventual settlement. Dr. Mitchell, who, particularly as a lecturer at the school at Walhalla, came in direct contact with the CO's,

ference, New York City, N. Y.—and the mimeographed copies of the C.F.C.C. *Newsletter,* issued by the Canadian Fellowship for Cooperative Communities, Toronto, Ont.

[5] *The Community Broadsheet, op. cit.*

offered his place in Georgia as a convenient opportunity for such settlement. From 1945 on CO's, released from the camps or from prisons, began to drift to Macedonia, to "have a look at it." Of the many who came only a few stayed on. Several of the group of fifteen in residence by February, 1947, felt that their stay was only temporary.

Analysis of past and present cooperative communities shows that effective selection is one of the main factors in their success.[6] Dr. Mitchell, being against "choosey selection of membership," apparently believed that the best way to proceed was "to let the community grow freely, naturally." [7] Consequently, at Macedonia everybody was welcome. No rules or regulations existed. Those who took up residence were simply expected to take part in the activities.

But a certain "latent" principle of selection may be discovered in the circumstances under which the members were attracted to Macedonia. They had known each other in the CPS camps, they shared the same ideal, and had suffered for it. By showing up at Macedonia, they meant to bear testimony to their sense of mission and to the high standards of performance its carrying out required. The hard test of reality was too exacting for the many who quit. And this in itself was one form of selection.

At the time of the study, the group consisted of three bachelors, six married couples, and their five children. The members lived in the houses formerly occupied by the local group. There were three such dwellings, two of them housing five of the couples, and a smaller one, inhabited by one couple. A new house had been built for an old lady who did not participate in the work of the community. One of the bachelors

[6] Cf. Henrik F. Infield and Ernest Dichter: "Who is Fit for Cooperative Farming," *Applied Anthropology,* II (January-March 1943).

[7] Notes made from the talk of Morris Mitchell in Macedonia, *Viewpoint,* Ft. Steilacoom, Washington, December, 1943 (Mimeographed.)

was quartered in this house; the other two roomed together in the larger of the two houses occupied by two couples.

The total property, owned by Dr. Mitchell, included 970 acres, of which 725 acres were protected woods, 135 acres wood pasture, 57 acres pasture, and only 9 acres crop land. In addition, the community had the free use of 200 acres of timber land. The dairy was stocked by seventeen milking and ten dry cows, twenty-five heifers and calves, and two bulls. Fifty-seven chickens represented the remainder of an abandoned attempt at poultry raising. There was also the sawmill and the apiary; and a woodwork shop had recently been started.

In theory, the community practiced cooperative production and individual consumption. The members were to receive going wages for work, and to live in individual households. At the end of the year, any resulting profit was to be shared.

But things were not going well. Since there was no formal binding, people would leave, for bad or for good reasons, but in any case without regard for economic necessities. Activities depending on their participation would simply have to be dropped. All land and buildings were held as collateral by the Farmers' Home Administration (successor to the FSA), and by the Federal Land Bank. The group was in debt to the grocer, hardware store, and auto repair man, as well as to its own members, some of whom had claims amounting to several hundred dollars. Income was too low even for mere subsistence. There seemed to be no sense in keeping up the pretense of wage credits which were obviously irredeemable, and of profits which appeared illusory. In practice, each worked according to his abilities and received as much, or rather as little, for the satisfaction of his needs as the available resources would yield. Thus, the kind of cooperation adopted was just about as extreme as that we saw practiced at Matador, although for different reasons.

The Studies and Their Results

The study of Matador was made in July, 1946, that of Macedonia in February, 1947. In both studies, as we mentioned, the same techniques of investigation were used.

The biographical group interview was conducted in exactly the same manner. It was opened, in both cases, by a few explanatory remarks and by a reading of the section related to life histories, from the article "Who Is Fit For Cooperative Farming?" [8] At Matador, fourteen of the fifteen members participated in the group interview, and related their life histories in one prolonged afternoon session. At Macedonia, the sessions were extended over three evenings, and all but three members, who were absent from the group, took part.

The data obtained have been summed up, in Table I, for Matador, and in Table II, for Macedonia. In both cases, the questions were formulated in the same manner and covered items enumerated in the tabulation. Once it was felt that the scope of information desired was set in the minds of the participants, no interruption was made, even if it meant neglect of certain aspects of the life history. Some incompleteness will, therefore, be found, particularly in Table I.

A "job analysis" preceded, in both groups, the formulation of questions for the sociometric test.[9] At Matador, this was done with the help of an official of the Department of Cooperation who had been in close touch with the group from its very beginning, at Macedonia, with the assistance of the acting manager of the group. The meaning of the eight questions decided upon was in both instances essentially the same. A slight change in the wording was necessary chiefly because

[8] Infield & Dichter, op. cit.

[9] For basic information on the sociometric test see: J. L. Moreno, Who Shall Survive? (New York: Beacon House, 1939), Helen H. Jennings, Leadership and Isolation (New York: Longmans, Green, 1943), and: Sociometry, A Journal of Inter-Personal Relations, passim.

TABLE I

Data Obtained from Biographical Group Interview at Matador

	Names of Members	*Number*
A. Background and Personal Data		
1—Origin		
Born outside Canada	E, I	2
Canadian of foreign born parents ⎰ Anglo-Saxon	C, D*, F*, K, N	5
⎱ Other	G*	1
Canadian of Canadian parents	A, B*, H*, J*, L*, M	6
*Born in Saskatchewan		
		14
2—Father's Occupation		
Farmer	A, B, C, D, E, F, G, H, J, K, L	10
Tailor	I	1
Ferry Man	G	1
Railroad Worker	M, N	2
		14
3—Size of Parental Family		
3-4 Siblings	I	1
More than 5 Siblings	A, C, D, E, F, G, H, J, L	9
		10
4—Sex		
Male	All Members	15
Female	None	0
		15
5—Age		
Between 25-30	A, B, D, F, J, L	6
Between 30-36	C, E, G, H, I, K, M, N	8
		14
6—Education		
Public School up to Eighth	E, D, G, H, L, N	6
1-2 years High School	A, C, K, M	4
High School Completed	B, I	2
University	J	1
Other Special Studies	F	1
		14
7—Religion		
Protestant	A, B, C, D, E, G, I, J, K, M, N	11
Roman Catholic	H, L	2
Greek Orthodox	F	1
		14

TABLE I—*Continued*

	Names of Members	Number
8—Marital Status		
Single	A, B, C, D, F, H, G, I, J, L	10
Married	E, K, M, N	4
		14
B. Life Experiences		
1—Occupational Changes		
1-2 Changes	B, C, E, L, M, N	6
3-5 Changes	A, G, H, J, K	5
7-8 Changes	D, I	2
14 Changes	F	1
		14
2—Military Service		
Army	D, E, G, H, I, K, L, N	8
Air Force	A, C, F, M	4
Navy	B, J	2
		14
3—Length of Service		
3 Years	D	1
3-4 Years	A, B, F, H, I, J, K, L, M, N	11
More than 4 Years	C, E	2
		14
C. Attitudes		
1—To Cooperative Farming		
Predominantly Economic Motives	E, F, H, I, J, K, L	7
Some Emphasis on Ideology	A, B, C, D, G, M	6
		13
2—To Marriage		
Favorable in General	C, F, H, I, L	5
Considered Only Girl Willing to Join	D, G, J	3
		8

of the different kind of leisure time activities preferred, or possible, in each setting.

The Matador questionnaire read as follows: Whom do you choose for: 1) working together? 2) bunking next to you? 3) sitting next to you at meals? 4) going hunting or fishing? 5) going together to town on free evenings? 6) visiting sports and social affairs? 7) talking over problems of the farm? 8) talking over personal affairs?

TABLE II

Data Obtained from Biographical Group Interview at Macedonia

	Names of Members	*Number*
A. Background and Personal Data		
1—Origin		
Born outside U.S.A.	F	1
U.S. of Foreign Born (Anglo-Saxon)	C	1
U.S. of U.S. Parents	B*, D*, E, G, H*, I, J, K, L, M	10
*Born in the South (but not in Georgia)		12
2—Father's Occupation		
Farmer	D, E, I, J, M	5
Profession	B, C, F, H	4
Businessman	G, K, L	3
		12
3—Size of Parental Family		
2-3 Siblings	B, D, E, F, G, H, I, K	8
4 or more Siblings	C, J, L, M	4
		12
4—Sex		
Male	A, C, F, G, I, J, K, M, O	9
Female	B, D, E, H, L, N	6
		15
5—Age		
Between 24-30	B, E, F, G, J, K, L	7
Between 30-35	C, D, H, I, M	5
		12
6—Education		
High School Only	M	1
College or Equivalent	B, C, D, E, F, G, H, I, J, K	10
Special Studies	L	1
		12
7—Religion		
Methodist	D, E, G, J	4
Presbyterian	F, B	2
Fundamentalist	L	1
Episcopalian	C, K	2
Church of Brethren	J	1
Lutheran	M	1
Quaker	H	1
		12

TABLE II—*Continued*

	Names of Members	Number
8—Marital Status		
Single { Male	G, J, K	3
{ Female		0
Married	A & B; C & D; E & F	
	H & I; L & M; N & O	12
		15

B. Life Experiences

	Names of Members	Number
1—Occupational Changes		
1-2 Changes	B, F, J, K	4
3-5 Changes	D, E, G, H, L, M	6
7 Changes	C, I	2
		12

	Names of Members	Number
2—War Experiences		
CPS Camps Only	G, J, K, M	4
CPS Camps plus Prison	C, F, I	3
		7

	Names of Members	Number
3—Length of Detention		
3 Years	G	1
3-4 Years	F, I	2
More than 4 Years	C, J, K, M	4
		7

C. Attitudes

	Names of Members	Number
1—To Cooperative Community		
Predominantly Ideological	C, D, E, F, G, J, K, M	8
Some Emphasis on Economic Motives	B, H, L	3
Doubtful (for Professional Reasons)	I	1
		12

The questions, or "criteria," on which the members of Macedonia were asked to make their choices, were: Whom do you choose for: 1) working together? 2) having as neighbor? 3) visiting at meals and for an evening? 4) going for walks? 5) going on trips 6) talking over personal problems? 7) talking over business problems of the community? 8) exchanging ideas of a general nature?

The number of choices allowed on each criterion was, in

both cases, the same, namely four. The members were left free to express only preferences they really felt; some of them did not utilize the full range of choices.[10]

About the same length of time, an hour and a half, was required for the application of the sociometric test in each case. At Matador, two members were absent, and one refused to participate because, as he said, "they are all welcome," meaning that he had no preferences. At Macedonia, all but the absent members registered their choices. Thus, the number of participants was in both groups exactly the same, twelve.

The results of the sociometric tests are recorded in Table III, for Matador, and in Table IV, for Macedonia. For technical reasons, we present only the totals needed for the computation of the raw scores of social status.[11]

TABLE III

Choices Received by Each Person On All Criteria
in the Sociometric Test at Matador

	A	B	C	D	E	F	G	H	I	J	K	L	M	N	O
1) Working	8	6	5	4	4	2	2	1	2	1	0	2	1	4	3
2) Bunking	4	4	4	4	2	3	1	0	2	1	1	1	0	0	2
3) Meals	5	0	5	3	5	0	3	0	0	1	1	2	3	1	1
4) Hunting	2	5	5	4	2	3	0	1	1	4	0	1	0	2	3
5) Town Visits .	3	7	9	4	2	2	0	2	0	3	0	2	1	1	3
6) Sports	6	8	7	4	0	1	1	3	1	4	0	1	2	0	3
7) Farm Problems ...	4	3	7	2	3	2	4	0	2	4	2	0	3	4	1
8) Personal Problems ...	5	4	4	3	1	2	1	2	3	4	1	1	4	3	0
Social Status Raw Scores	37	37	46	28	19	15	12	9	11	22	5	10	14	15	16

[10] Urie Bronfenbrenner: "A Constant Frame of Reference for Sociometric Research," *Sociometry*, VII (February, 1944), (Part II), p. 44.

[11] To prevent any possible embarrassment, letters in alphabetic order which have no reference to the names themselves are used to designate the members in all tabulations.

TABLE IV

Choices Received by Each Person On All Criteria
in the Sociometric Test at Macedonia

	A	B	C	D	E	F	G	H	I	J	K	L	M	N	O
1) Working ...	3	3	6	0	4	7	0	1	0	7	0	3	1	4	5
2) Neighbor ...	2	2	6	1	3	5	1	0	0	5	1	4	1	4	4
3) Visiting	5	2	5	2	4	6	0	2	3	1	2	3	2	3	2
4) Walks	6	1	4	2	3	3	0	1	3	2	2	2	2	1	3
5) Trips	6	3	6	1	6	6	0	1	1	2	1	2	1	0	2
6) Personal Problems ...	5	3	4	1	8	5	1	3	1	2	1	1	1	2	2
7) Business ...	5	6	10	1	2	9	2	2	1	3	0	1	1	0	2
8) General Ideas	7	2	5	1	4	5	0	2	3	4	1	1	2	1	2
Social Status Raw Scores	39	22	46	9	34	46	4	12	12	26	8	17	11	15	22

The Biographical Group Interview

The history of the modern cooperative community is a relatively short one. Hardly any scientific analysis has yet been made of the factors contributing to its success, or failure. Available material seems to indicate that its fate is largely dependent upon the ability of members "to get along with each other." If this is true, a basis for predicting success could be found by determining the amount of ability to work together and to live together present in a given community.

The assumption that certain data in the members' life history are indicative of fitness for cooperative farming has helped set the pattern for our group interview.[12] Looking at Table I and Table II, we shall find that virtually all the information obtained can be related to capacity either for work or for cooperation. This information is not of the kind which would make possible utmost mathematical precision; but it is precise enough to allow a quantified estimate of the two groups' comparative chances for succeeding in their like undertaking.

[12] Infield and Dichter, *op. cit.*

As data related to the capacity for work we propose to consider those listed under the headings of "Father's Occupation," "Sex," and "Age"; and those recorded under "Size of Family," "Education," "Marital Status," and "Life Experiences" as having a bearing on ability to cooperate. In addition, we shall deal briefly with the information contained under "Origin," "Religion," and "Attitudes."

As to work, the one dominating fact to remember is that both groups are land settlements. Early familiarity with farming cannot but be a definite asset in such settlements. It is in this connection that the father's occupation becomes highly significant. As our tables show, of fourteen members at Matador, only four give as their father's occupation any other than farming, and it is in this case that of laborer. At Macedonia, out of twelve members seven state to have come from professional (4), or business (3) environments, and only five from rural backgrounds. It is obvious that Matador commands a far superior farm experience than Macedonia.

A work potential superior to that of Macedonia is inherent also in the sex composition of Matador. At the time of the study, all fifteen members of Matador were men; at Macedonia, there were nine men and six women. In the first group, the whole membership could thus devote itself to agricultural production; in the second, two fifths of the available labor force had to be spent on "unproductive" tasks.

The only approximate equality can be found in the age range of the two groups, although the Matador membership as a whole ranks higher. Six of the Matador members were between twenty-four and thirty, and eight between thirty and thirty-six. At Macedonia, the majority, seven, were between twenty-four and thirty, and only five between thirty and thirty-five.

As to cooperative ability, we want to consider first information recorded under "Size of Family." It may be assumed that the chance of a member to have developed habits of give-and-

take, so important in cooperative living, will have been the
better the larger the family in which he grew up. Of the ten
members at Matador who have offered information on this
point, nine have grown up in families of five or more siblings,
and one in a family of four siblings. Of the Macedonia group,
on the other hand, eight came from families of two or three,
and only four from families of four or more siblings. Thus it
would seem as if here again the Matador group would score
over Macedonia.

But the picture changes when we look at "Education." Only
two of the Matador group completed high school, and only
two others continued beyond, one going for a time to the
University, and one receiving some special training. Against
this, of the ten members of Macedonia who went to college,
seven—four of the women and three of the men—obtained a
degree; one other had had special training after high school,
and there was only one whose formal education had ended
with high school. Experience shows that cooperation, to be
successful, requires a relatively high standard of education.
The one instance of cooperative community which now is generally accepted as most successful—the Israeli *kvutza*—
counted among its founders fifty-six per cent with higher education.[13] Thus, as far as this factor goes, Macedonia would
seem to have a definite advantage over Matador.

This appears to be true also with regard to the marital status
of members in the two groups. Cooperative settlements tend
to be more stable the better the two sexes are matched. Most
favorable is a situation where the number of men and women
is about equal at the start, and where after a time only a
negligible number of members is left single. Of the twelve
members at Macedonia only three were single, and all the rest
were married couples. At Matador, only men were members

---

[13] The levels of education are compared here in the sense of a rough
approximation, without regard to variations in content, scope, and so
on, in the different settings.

at the time of the study. Four of them were married, and there was talk about making the women full fledged members later on. Even so, eleven of the men were bachelors, and although three of these declared that they would marry only girls willing to join, it was obvious that Matador was exposed to the disintegrating possibilities of out-of-group mating to a much higher degree than Macedonia.

More evenly matched were the two groups with regard to the influence their life experiences may have had on their cooperative way of living. Statistically, at least, there seems to be about an equal amount of occupational changes experienced by members of both, particularly if we disregard the extreme case of fourteen changes at Matador. In any case, since both counts have been taken on the basis of the same, admittedly crude, measurement, we can be sure that we are not comparing disparate units. By multiplying the number of changes in each group with the upper limit of the range, we obtain 8 plus 30 plus 14, or 52 for Macedonia, and 12 plus 15 plus 16, or 53 for Matador. Divided by the number of members concerned, this would give Macedonia a slightly better average score per member (4.3) than Matador (4.0).

It may seem more difficult to find a way of comparing the war experiences of the two groups. In this respect they appear diametrically opposed. The attitude in the case of Matador is one of voluntary acceptance of armed service, in complete conformity with existing mores; in the case of Macedonia, it is conscientious objection in the face of strong social disapproval, a radically non-conforming attitude. The socio-psychological implications of service in the army on the one hand, and of conscientious objection, on the other, would seem to make any comparison impossible. It is only when we consider the strictly sociological aspects of the apparently differing experiences that we find them not so divergent after all. For the dominant single restrictive factor in both, the CPS camp as well as the Army, has been enforced submission to imposed discipline; and the dominant relieving factor in both cases has been the sense of being one of a team. In terms of the dominant mode of interpersonal relations prevailing both in the CPS camp and the Army, the experience, and its potential influence upon cooperative ability would then appear to have been not at all unlike. If we accept this, any difference will have to be found in quantitative factors, in the length of service or detention, and in the number of participants.

It would seem that on both these scores Matador ranks over Macedonia. For the time spent in the armed forces was longer, on the whole, than that spent in the CPS camps. And since all the members of Matador were veterans, while only the men but not the women of Macedonia went to CPS camps,[14] the amount of direct participation must obviously be larger in the first case.

There was, however, one aspect of the situation which, in our estimate, favored Macedonia. This was its nonconformist background. The sense of fellowship in the CPS camps must have been more intensive than in any military unit. The CO's feeling of belonging together was not the product of a more or less mechanical assignment; they had been adherents of the same movement before they ever met in the camps. Here, the hostility of the outside world drove them even closer together; and the sense of marginality, persisting even after release, made their need for each other a stronger force of coherence than that which may grow out of any occasional nostalgia for "the buddies." We may, therefore, be justified in concluding that, as far as cooperative living goes, the experience in the CPS camps has been more conducive to cooperation than that in units of the armed forces.

Of some significance in evaluating their comparative chances of success may be—in addition to the above—the fact that seven of the members were born in the same province of

[14] Only two of the women shared in this experience directly, by becoming employees in CPS camps.

Canada where Matador is located; on the other hand, of the members of Macedonia, three were born in the South, but none in Georgia.[15] Since relations to the neighborhood are important for a modern cooperative community, it is obvious that Macedonia would have to make more strenuous efforts than Matador to become well accepted.

A lesser factor is, probably, the difference in the denominational make-up of the two groups. Religious observance plays only a negligible role in both.

Considering all information obtained from the group interview, we may then sum up as follows: As to work potential, Matador scores over Macedonia in two out of three points—in familiarity with farming and in sex composition, and is about equal in one, age range. But if compared as to cooperative ability, Macedonia appears to be superior in three out of four points—education, marital status, life experiences.

Accordingly, we may expect that in actual performance Matator will surpass Macedonia in work achievements, but will experience more difficulties in interpersonal relations. The attitude towards cooperative farming as expressed by the respective members[16] can be taken as an added bit of information supporting this conclusion. Although six of the Matador members were aware of the ideological aspects of the cooperative community, all thirteen who offered statements on the subject professed to have joined mainly for economic reasons. In Macedonia, on the other hand, all members stressed the ideological motives for joining, and only three mentioned, in addition, economic considerations.

The Sociometric Test

We shall find more precise confirmation of our conclusion in the results obtained from sociometric tests. The comparison of these results has been considerably facilitated by

---

[15] See Table I and II, under "Origin."
[16] See Table I and II, under "Attitudes."

the existence of a "constant frame of reference for sociometric research." [17]

Bronfenbrenner's contribution would have been valuable had he done no more than develop a mathematically sound method for identifying the statistical significance of sociometric data. But he has earned the gratitude of workers in the field by simplifying his refined sociometric techniques to the point where they can be used with relative ease, and "without too great a loss in the accuracy of the results." [18] In two tables—one for "Critical Raw Status Score Values" and the other for "Critical Raw Mutual Choice Values" [19]—he offers standardized scores valid for "diverse sociometric situations."

Unfortunately, we are prevented from making full use of the tables because of the special field of our investigation. As we have seen, the application of the sociometric test to cooperative community required a larger than the usual number of criteria. Legitimately considering only the needs of "current sociometric practice," Bronfenbrenner calculates his standard scores for no more than three criteria, while the results of our tests were based on eight criteria. It is true that Bronfenbrenner himself foresees the need for added computations related to more elaborate tests. [20] But as long as such computations are pending, we are faced with the choice of either undertaking elaborate calculations of our own, or of reducing the number of criteria to three.

Deciding for the sake of expediency on the latter course, we proceed then to select from both tests three criteria which can be considered as equivalent. The task is made easier because of the fact that two of the criteria are identical in both questionnaires: "working together," and "talking over personal problems." It is the selection of the third criterion which offers

[17] Bronfenbrenner, *op. cit.,* p. 44.
[18] *Ibid.,* p. 66.
[19] *Ibid.,* Tables XVII and XVIII, p. 66 f.
[20] *Ibid.,* p. 67, note 10.

some difficulty. We may eliminate the criterion of "talking over problems of the farm" on the one hand, and "talking over business problems" on the other, on the ground that they are covering aspects similar to "working together." There remain then five criteria from which to choose. Of these, there are three on each side which are related to leisure time activities. A look at the totals of choices registered on each of these criteria shows that "visiting sports and social affairs" (Matador) and "visiting at meals and for an evening" (Macedonia) receive almost the same amount of attention (41 and 42 choices respectively). We obtain thus the criterion to be added, on each side, to the two named above. In Tables V and VI the reader will find the choices registered on the three selected criteria, together with the raw scores of social status.

Looking at the raw status scores based on choices received, in both tables, we find the highest scores at Matador to be 19 (A) and 18 (B), and the lowest 1 (K); at Macedonia, the highest score is 18 (F), and the lowest 1 (G). To obtain a clearer picture of respective stratification, we consult Bronfenbrenner's Table XVII. We are using results based on three criteria, and four choices. Accordingly, our expected value is 12, the lower limit 5 and the upper 18. This enables us to divide all scores into four brackets: (1) the top—or "star" bracket, including all those with scores of 18 or more; (2) the upper bracket, those with scores of more than 12 but less than 18; (3) the middle bracket, those with scores of less than 12 but more than 5; and (4) the lowest bracket or "neglectees," those with a score of 5 and less. In this way we obtain the following table of comparative stratification:

The first thing that strikes us about this table is that the number of "stars" is much smaller than that of "neglectees." There are two stars at Matador and one at Macedonia; but there are five neglectees on each side. This tends to confirm, and to extend to adult groups, findings of previous research, particularly of Bronfenbrenner who found that "in classroom

TABLE V

Frequency Distribution of Sociometric Choice by Three Criteria, *Matador*.

Subjects	A	B	C	D	E	F	G	H	I	J	K	L	M	N	O
A........		122	211		400			004				040	003		330
B........	200		111					040		023		400	002		330
C........	223	111		332	400								004	402	040
D........	030	200	300		004		140	020		010	001		003		
E........	112	330	220						443	001					
F........	121	244	033		300					412					
G........	421								004					300	200
H........															
I........	300		400										200		
J........	400	233	020	340	100								010	001	
K........		030	021	202		104				003			010	100	
L........	103	020	030			201			302	010				400	
M........															
N........															
O........	010	020		043		010	400	132				201		304	

	A	B	C	D	E	F	G	H	I	J	K	L	M	N	O
Totals.........															
Working.........	8	6	5	4	4	2	2	1	2	1	0	2	1	5	3
Sports.........	6	8	7	4	0	1	1	3	1	4	0	1	2	0	3
Personal Prob.	5	4	4	3	1	2	1	2	3	4	1	1	4	3	0
Social Status Raw Scores....	19	18	16	11	5	5	4	6	6	9	1	4	7	8	6

TABLE VI

Frequency Distribution of Sociometric Choice by Three Criteria, *Macedonia*.

Subjects	A	B	C	D	E	F	G	H	I	J	K	L	M	N	O
A.															
B.	102		220	030	001					300	040	400		013	112
C.		410		001						244				020	200
D.						333								320	200
E.	020		101									440			
F.	013		013		101			232	040	300				400	400
G.	030		322		004	101		030	020	200	003		040		
H.			300	010	232	111				202		100		300	
I.		003	020		024	040	004	001				030			
J.	300		240		002	112									101
K.	111	322	203		003	400						040	030	004	
L.	002	403			140	400				200			201	300	020
M.	012				433	344		004	030						100
N.										200	020	001			
O.															

	A	B	C	D	E	F	G	H	I	J	K	L	M	N	O
Totals Working	3	3	6	0	4	7	0	1	0	7	0	3	1	4	5
Visiting	5	2	5	2	4	6	0	2	3	1	2	3	2	3	2
Personal Prob.	5	3	4	1	8	5	1	3	1	2	1	1	1	2	2
Social Status Raw Scores	13	8	15	3	16	18	1	6	4	10	3	7	4	9	9

TABLE VII

The social status raw scores ordered according to brackets

	Matador	Macedonia
	A(19)	
"Stars"	B(18)	F(18)
Upper bracket	C(16)	E(16)
		C(15)
		A(13)
Middle bracket	D(11)	J(10)
	J(9)	N(9)
	N(8)	O(9)
	M(7)	B(8)
	H(6)	L(7)
	I(6)	H(6)
	O(6)	
"Neglectees"	E(5)	I(4)
	F(5)	M(4)
	G(4)	D(3)
	L(4)	K(3)
	K(1)	G(1)

groups the proportion of overlooked or rejected children is greater than the proportion of children who are exceptionally popular and acceptable." [21]

Comparing their stratification as a whole, we find a marked difference between the two groups. At Matador, there are two stars, one member in the upper and seven members in the middle bracket, and five are "neglectees"; at Macedonia, the corresponding numbers are one, three, six and five. This would indicate that the distribution is more even at Macedonia than at Matador, and consequently, that the ties growing out of intimate acquaintances are weaker in the latter case. [22]

[21] *Ibid.*, p. 58.

[22] Cf. Bronfenbrenner's statement with respect to "intensity of star-dom." *Ibid.*, p. 74: "As members of groups become better acquainted,

But it is from the consideration of mutual choices that we gain a clear picture of the degree of group coherence attained in each of the two groups. The graphic presentation of the two group structures (see Tables VIII and IX of Sociograms), based on mutual choices on the three criteria, brings into focus their comparative strength and weakness. A distinction of the Macedonia sociogram, as compared with that of Matador, is the division of the "target" into halves. The vertical line, in accordance with current practice of sociogram construction[23] mechanically segregates the two sexes present at Macedonia; no such dividing line is needed for Matador.

But the cleavage has more than a mechanical significance. Looking closer, we notice that of the seven mutual choices on the work criterion (straight lines) registered at Macedonia, six are evenly divided between men (3) and women (3); only one links the two sexes (E x F). E and F who are married to each other (see Table II) are, in turn, "chained" with all whose choices on this criterion are reciprocated: E with H, L, and B; F with C and J. Thus the link between E and F would indicate that these two occupy a "key" position in the network of work relationships.

A more frequent crossing of the dividing line occurs on the "visiting" criterion (broken lines). Here, out of nine mutual choices, six are again distributed evenly on each side, but the remaining three cut the "sex" line. To evaluate this fact properly, we have to consider that "visiting" means actually "wanting to be together with somebody in one's spare time": choices registered on this criterion are an index of what we might call "personal intimacy." Couples, since they live together, were asked to disregard their mates under this heading. The fact

the amount of attention commanded by persons who are centers of attraction decreases slightly."

[23] Urie Bronfenbrenner, "The Graphic Presentation of Sociometric Data," *Sociometry,* VII (August, 1944); Mary L. Northway, "A Method for Depicting Social Relationships by Sociometric Testing," *Sociometry,* III (April, 1940).

that the number of "intersex" choices equals that between members of each of the two sexes would indicate, therefore, that sex differences can be hardly any barrier to personal intimacy at Macedonia.

Equally revealing is the manner in which the lines are drawn representing mutual choices on the criterion of "talking over personal problems" (dotted lines), a criterion which we take as indicative of mutual confidence. Of eight such lines, four indicate "intersex" choices (E x F; D x C; H x I; L x M), all between wives and husbands. There can be hardly anything surprising in that. Still, it would seem to indicate a great amount of mutual confidence between the married.[24] Of greater interest is the observation that out of the remaining four choices, three are spent on the male side (F x C; C x J; J x G), and that in the form of a "chain"; but only one on the female side. The degree of mutual confidence[25] between men appears to be much higher than that between women in this group.

Next to attract our attention is the difference between the raw scores of social status (see Table VII), and the mutual choice scores as revealed in the sociograms. Looking first at Matador, we find on the first criterion a "chain" between A, B, and C, extending to D, G, and O, and E and I. But the "key" in this "chain" are neither A nor B, who rank as "stars" on the raw social status score, but C whose all four choices on this criterion are reciprocated. In fact, he outranks, with a mutual choice score of 8, both B, with 7, and A with 6, and emerges as the "key" individual in the total group structure of Matador.

No such startling difference between the two scores can be noticed within the higher brackets in the Macedonia socio-

[24] B too has chosen A (husband), but because of his absence mutuality remains undetermined.

[25] Interchoice intimacy on a personal basis, in this sense, has been differentiated from interchoice on a less personal work basis, by Helen H. Jennings, in "Sociometric Differentiation of the Psychegroup and the Sociogroup," *Sociometry Monographs,* No. 14.

TABLE VIII

SOCIOGRAM OF MUTUAL CHOICES
ON THREE CRITERIA: MATADOR

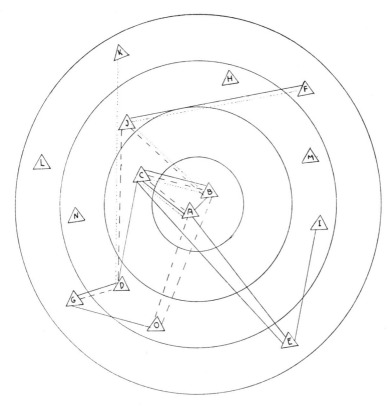

Working: Straight Lines
Sports: Broken Lines
Personal Problems: Dotted Lines

TABLE IX

SOCIOGRAM OF MUTUAL CHOICES
ON THREE CRITERIA: MACEDONIA

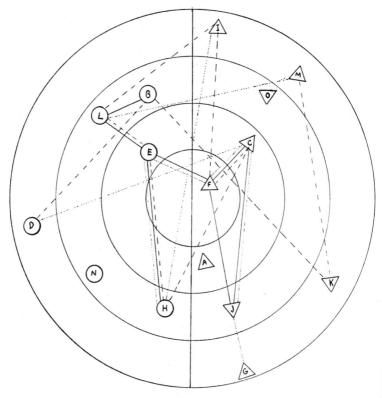

Triangles: Male

Circles: Female

Working: Straight Lines

Visiting: Broken Lines

Personal Problems: Dotted Lines

gram. F and E who rank highest on the raw social status score (18 and 16) maintain, with 8 and 7, their respective rank on the mutual choice score. Thus, while in the all male group of Matador the key position is held by one individual, the same position is occupied in the bisexual Macedonia group, appropriately enough, by a couple.

On the other end of the scale we find, looking at Matador, that the "neglectees," F and G (see Table VII) not only reverse their respective positions (G with 3 and F with 2 choices reciprocated), but also outrank I, who, in spite of being a member of the "middle bracket" scores only 1 on the mutual choice score.

More drastic is the change in status occurring in the lower brackets on the Macedonia side. H, who comes close to being a "neglectee," moves on the mutual choice score up to fourth place.

All in all, it can be seen that the mutual choice score helps to fix more precisely than the raw social status score a mem-

TABLE X

Mutual Choice Scores on Three Criteria for Matador and Macedonia.

MATADOR

Member	Work	Sports	Personal	Total
C	4	2	2	8
B	2	3	2	7
A	3	2	1	6
D	2	2	1	5
J	1	2	2	5
E	3	0	0	3
G	2	1	0	3
O	1	2	0	3
F	1	0	1	2
I	1	0	0	1
K	0	0	1	1
L	0	0	0	0
Total	20	14	10	44

MACEDONIA

Member	Work	Sports	Personal	Total
F	3	3	2	8
E	3	2	2	7
C	2	1	3	6
H	1	2	2	5
L	2	2	1	5
J	2	0	2	4
B	1	2	0	3
I	0	2	1	3
D	0	1	1	2
K	0	2	0	2
M	0	1	1	2
G	0	0	1	1
Total	14	18	16	48

ber's position within the group's social structure. Consequently, we can expect to receive a true picture of their relative degree of group coherence by comparing the two groups' total mutual choice configurations. To facilitate such comparison, we tabulate the mutual choices for each member, as registered on each criterion.

Consultation of these tables reveals some deviation from findings of former sociometric research. In his classroom investigation, for instance, Bronfenbrenner found "that children who are accepted on the basis of one criterion are likely to be accepted on the basis of other criteria as well, particularly when the activities or situations are simultaneous or overlapping." [26] There can be no doubt that the activities symbolized by our three criteria are "simultaneous and overlapping." Nevertheless, our results seem to bear out Bronfenbrenner's findings on this point only as far as the highest brackets are concerned. On the lower brackets our results differ. E (Matador), for in-

[26] Bronfenbrenner, "A Constant Frame of Reference for Sociometric Research," *op. cit.*, p. 47. (However, Bronfenbrenner admits that, "considerable variation may occur in individual instances." p. 71.)

stance, who scores 3 on one criterion does not score at all on the other two criteria. On the other hand, I, D and M (Macedonia) who do not score on one criterion, score on both of the other two criteria.

This deviation assumes significance when the particular criteria are considered. On the first, the "work" criterion, at Matador, only two do not score at all, K and L; but there are five, E, F, I, K and L, who do not score on the second; and five again, E, G, I, L and O, who do not score on the third criterion. At Macedonia, however, while there are five, D, G, I, K and M, who do not score on the first criterion there are only two G and J, who do not score on the second, and two again, B and K, who do not score on the third criterion. Holding the first against the two other criteria the figures are exactly reversed in the two groups.

If we merely looked at the total mutual choice scores in the two groups, we would only gather that both must be about equally well integrated. Consulting Bronfenbrenner's Table XVIII[27] we find that on three criteria and four choices the lower limit is 16, the upper 36. A look at our table shows that the total of Matador is 44, and of Macedonia 48; which, since all reciprocated choices are counted, has to be divided by two, thus yielding 22 and 24 respectively. The amount of group coherence, taken as a whole on this measure, would appear to be about equally distant from the lower and upper limit in both groups.

A marked difference between the two groups becomes apparent however, once we break down the totals into component parts. The comparative figures on each of the three criteria are, for Matador: 10, 7, 5; and for Macedonia: 7, 9, 8. Consulting again Bronfenbrenner's Table XVIII, but this time on one criterion, we find the lower limit given as 3, and the upper as 15. Applied to our figures, this would mean that,

[27] *Ibid.*, p. 68. But note criticism in Joan H. Criswell, "Foundations of Sociometric Measurement," *Sociometry*, IX (February, 1946), p. 11.

while Matador is much closer to the upper limit on the first criterion, it is closest to the lower limit on the third, and somewhere in between on the second criterion. Against this, Macedonia is closest to the lower limit on the first criterion, moves closest to the upper limit on the second,[28] and less close on the third criterion.

In other words, while Matador is quite well teamed for work (first criterion), Macedonia, for one reason or another, appears to be weakest on this point. But while at Matador there is only fair mutuality in leisure time activities (second criterion), and only relatively little mutual personal confidence, Macedonia, in spite of sex cleavage, shows a far higher degree of mutual confidence among its members (third criterion), and an almost optimal amount of mutuality in leisure time activities. Thus, as we see, the results of the sociometric tests bear out, only in a more precise fashion, our conclusion based on the data from the biographical group interview. Compared with Macedonia, Matador is better geared for work, as apparent from their better integrated sociometric structure along work lines. This has an obvious bearing on each of the two groups' chances for survival. Should Macedonia's living standards become too low for continuous existence, the intimacy between its members would prove no salvation. But Matador, as long as it maintains a sound economic basis, has a good chance of becoming better integrated along other lines in time. There is this difference, though: assuming that both would disband at the given stage, the members of Matador would part without having formed any deeper personal attachments; while the people of Macedonia would, most probably, remain friends beyond the conclusion of their experiment in common living.

[28] Remembering what we said about the latency of mutual choices on "visiting," the number of concealed mutual choices added to those manifested would bring Macedonia even closer to the upper limit on this criterion.

Theoretical Implications

The examination of our data on Matador and Macedonia has demonstrated that quantified group comparison is feasible. But there is one more point that has to be dealt with: the bearing of our study on the problem of cumulative research in sociology. A full length discussion of this aspect of our study would require more space than can be found in a single essay. We shall limit ourselves here to show merely that (1) our findings are sociologically relevant; that (2) the methods employed are valid beyond the specific subject of our investigation; and that (3) the use of these methods is apt to produce cumulative results in sociological research. In doing so, we shall confine ourselves to a statement of our argument, followed by a brief explanatory discussion.

The Sociological Relevance of the Findings

The modern cooperative community has generically much in common with society in general; hence our findings are sociologically relevant.

The opportunity offered to research by the modern cooperative community has first been called to the attention of social scientists in a note published by the American Sociological Review.[29] Since then, several studies[30] have shown that cooperative communities when fully developed, present on a smaller scale virtually all the essential aspects of society in general. These communities occupy a geographically defined territory of their own; they contain a more or less numerous population; their economic activities are geared to consump-

---

[29] Henrik F. Infield, "The Cooperative Community: A Note on a Potential New Field of Sociological Research," *American Sociological Review,* VII (December, 1942).

[30] Eaton, *op. cit.;* Henrik F. Infield, *Cooperative Living in Palestine* (New York: Henry Koosis & Co., 1948) and Henrik F. Infield, *Cooperative Communities at Work* (New York: The Dryden Press, 1945).

tion as well as to production; they face and have to solve problems of administration and government; they have to take care of their offspring, offer them education when they are young, and "careers" when they grow up; they have to pro-vide for their sick, invalid and aged; they have to make ar-rangements for recreational activities; and they have to satisfy, in one way or another, their members' spiritual needs. Conse-quently, all the social processes which occur in human society can be found, too, in the cooperative community. Neverthe-less, the cooperative community has many specific character-istics. Among these the most pronounced are: prevalence of face-to-face relations and cooperative way of living. It is only that these characteristics resolve themselves, in the sociological analysis, into differences of degree rather than of kind. For, face-to-face relations are by no means foreign to our own so-ciety; and cooperation is a mode of social relations without which no society whatever could hope to exist for long.

We may then be justified in concluding that, since coopera-tive community and society are generically alike, findings ob-tained from the study of one will be applicable to the under-standing of the other.

The Methods and Their Validity

The sociometric test and the biographical group interview are techniques of group investigation derived from the meth-odology of Systematic Sociology. This theoretical foundation makes them valid beyond the specific subject of our investi-gation.

The method followed in our analysis of Matador and Mace-donia is essentially that of Systematic Sociology. This school of sociology conceives society not as given and static, but rather as emerging and dynamic. Society, as von Wiese defines it, is "the total web of interaction called sociation." This "web of interaction" is divided into a large number of empirical social processes, or "actions which unite human beings with each other or cause them to separate."

In accordance with this concept, von Wiese following in the main Simmel, in Germany, and Ross and Cooley, in the U.S.A. undertook a "systematic classification of interhuman action patterns." This classification,[31] the most comprehensive to date, distinguishes three main groups of social processes: (1) processes of association, or those which "unite" or tend to draw people towards each other; (2) processes of dissociation, or those which "separate" or tend to drive people away from each other; and (3) mixed processes, which contain elements of both.

The merits of this systematization have not remained undisputed.[32] But, if we accept Poincaré's dictum that "One geometry cannot be more true than another: it can only be more convenient"[33]—the validity of this system should not be judged by its "originality" or even by its "truth," but rather, to use Poincaré's terms, by how well it "accords" with the "properties" of societal facts.[34] This again can best be gauged by the extent to which the system has proven to be productive in terms of fact finding.

There is hardly any doubt in this author's mind that research techniques like the sociometric test became possible only once the *specificum sociologicum* had been established by

[31] Leopold von Wiese, *Systematic Sociology—On the Basis of Beziehungslehre and Gebildelehre,* Adapted and Amplified by Howard Becker, (New York: John Wiley & Sons, 1932), p. 71; or Leopold von Wiese, *System der Allgemeinen Soziologie,* (Duncker & Humblot, 1933). (This is a second, revised edition of the system.)

[32] Pitirim Sorokin, *Contemporary Sociological Theories,* (New York: Harper and Brothers, 1928), pp. 495 ff. Howard Becker's reply to Sorokin's criticism, *Systematic Sociology, op. cit.,* pp. 43 ff.

[33] Henri Poincaré, *Science and Hypothesis,* authorized trans. G. B. Halstead (New York: The Science Press, 1905), p. 39.

[34] The reason given by Poincaré for accepting Euclidean geometry as the most convenient: "Because it accords sufficiently well with the

Systematic Sociology. For only then could it become apparent that mere adoption of techniques employed by other disciplines would not do. New techniques, capable of registering the dynamic properties of sociation, had to be invented. Of the several attempts to do so[35] the most consequential has proven to be so far that of sociometry.

Sociometry implies the acceptance of the *specificum sociologicum;* its initial methodological assumption is identical with that of Systematic Sociology.[36] Dr. Moreno's ingenuity lies not so much in the reduction of the classificatory terms of "association" and "disassociation" to their operational meaning of "attraction" and "repulsion."[37] It lies rather in the acceptance of the interpersonal relation as the elemental dynamic unit of any societal occurrence. This step followed methodologically from Systematic Sociology's definition of the social process as "actions which unite human beings with each other or cause them to separate." But it marks the significant advance from systematic theory to systematic fact find-

properties of natural solids, those bodies which our hands and eyes compare and with which we make our instruments of measure." *Ibid.*

[35] Kurt Lewin, *Principles of Topological Psychology,* trans. Fritz Heider and Grace M. Heider (New York: McGraw Hill, 1936); the same author's "Field Theory and Experiment in Social Psychology: Concepts and Methods," *American Journal of Sociology,* May, 1939; Ronald Lippitt, "An Experimental Study of Authoritarian and Democratic Group Atmospheres," *University of Iowa Studies,* February, 1940. An attempt to develop a device for the measurement of interaction was made by Eliot D. Chapple, "Measuring Human Relations: An Introduction to the Study of the Interaction of Individuals," *Genetic Psychology Monographs,* XXII (February, 1940). See particularly p. 34.

[36] Cf. Florian Znaniecki, "Sociometry and Sociology," *Sociometry,* VI (August, 1943); J. L. Moreno, "Contributions of Sociometry to Research Methodology in Sociology," *American Sociological Review,* XII (June, 1947).

[37] Any research taking as its starting point Systematic Sociology would have to reach, it appears, this point. Such, at least, was the experience of this author in his "test' application of von Wiese's system to the cooperative community in Palestine. The original manu-

ing. It is the dynamic character of the sociometric test which has made possible the detection of societal facts that went unnoticed before, and greater precision in the restatement of regularities formerly observed. As documented in the volumes of *Sociometry, A Journal of Interpersonal Relations*, the sociometric test was instrumental also in ensuring a remarkable continuity of research.[38]

The biographical group interview, although much cruder and used in subsidiary capacity, is in some respects similar to the sociometric test. It, too, aims at the group rather than at the individual; it transforms an essential part of the individual case study into group action; and it derives from the data obtained indices of group dynamics.

Both techniques show promise of becoming useful in control and prediction of success of cooperative groups; and both —the group interview much more than the sociometric test— stand in need of further improvement. This can be achieved only by more extensive, critical application. The use in comparative studies like the present one should prove particularly helpful in this respect.

The Possibility of Cumulative Research

By accepting interpersonal relation as the common unit of reference, our investigation helps indicate the feasibility of cumulative research in sociology.

In reviewing for the semicentennial issue of the *American Journal of Sociology* the history of sociological theory, Prof. Merton observes that "There are conspicuously few instances

script of *Cooperative Living in Palestine, op. cit.*, contained a section on *Beziehungslinien* used to design graphically attraction between group members. When it was found that sociometry had developed farther along the same lines, this section was later replaced by reference to the work of Dr. Moreno and his associates. pp. 142 ff.

[38] Cf. George A. Lundberg, "The Growth of Scientific Method," (particularly section B, "Atomic Research on Elementary Social Relations") *American Journal of Sociology*, (May, 1945).

of consecutive research which have cumulatively investigated a succession of hypotheses derived from a given theory."[39] The reason for this default may be found to lie not so much in an innate deficiency of sociology as a scientific discipline as in its practitioners' inability to free themselves from a preconceived idea about the requisites of such theory. These ideas appear to be compellingly influenced by the precedent of classical philosophy. Systems like those of Spinoza, Kant and Hegel tend to proceed from the fewest possible number of "given" theoretical principles towards a universal interpretation of the world in terms of utmost logical coherence and finality. The goal is here a rounded, all-inclusive picture; and the terms are those of a closed system.

Now, the one outstanding characteristic of society, accepted today by virtually all sociologists as "given," is that of change. No method can be derived from a closed system that would "accord" with this "property" of sociology's subject matter. Attempts to do so must, and have, ended in frustration and despair of all theory.

The alternative to a closed system, however, is not necessarily no system at all. Even philosophy has explored, for its own sake, the potentialities of the "open system"[40]—not without rejecting it in the end. But sociology, not aiming at transcendental finality, but rather more modestly, at the observable, may find it well suited to its needs.[41] As a matter of fact, it is as an "open system" that we can best describe the theoretical framework of Systematic Sociology.

It is only from the position of an "open system" that the

[39] Robert K. Merton, "Sociological Theory," *American Journal of Sociology*, L, (May, 1945), p. 472.

[40] Heinrich Rickert: "Vom System der Werte," *Logos*. IV, 1913. See in particular the section "Das Offene System," pp. 297-30.

[41] A consideration of such suitability is implied in H. Infield, "Die Fragmente des Novalis als Ausdrucksform eines Offenen Systems betrachtet" (Novalis' Fragments Interpreted in Terms of an Open System) (Unpublished Ph.D. Thesis, University of Vienna, 1925).

advance made by Dr. Moreno could be effected. For an "open system" is one that combines systematic coherence with unlimited expansibility. These characteristics correspond, on the action level, with purpose and spontaneity, the main determinants of human interaction.

It is this theoretical and methodological foundation which makes the interpersonal relation eminently "convenient" for sociological purposes. Its acceptance as a common unit of reference should naturally result in making sociological research cumulative.

Conclusion

Using data obtained in each case by means of the biographical group interview and the sociometric test, we have attempted a quantified comparison of the two cooperative communities, Matador and Macedonia. The one is a veterans' land settlement in Saskatchewan, Canada; the other, a pacifists' community in Georgia, U.S.A. Both were studied in the stage of formation.

The two groups were found to differ in respect to origin, background, and attitude; to be similar as to their socio-economic structure; and to have the same number of members.

The comparison of data obtained from the group interviews yielded the following scores: as to work potential, Matador ranks over Macedonia in two out of three points, familiarity with farming, and sex composition; and is about equal in one, age range. Against this, Macedonia surpasses Matador in three out of four points, cooperative ability, education, marital status, life experiences. These scores enabled us to predict tentatively that Matador may outdo Macedonia in work achievement, but may be expected to experience more difficulties in its social functions.

This prediction appeared to find support in the results of the sociometric test. Comparison of the mutual choice socio-

grams based on three criteria revealed that Matador scored
10 on the first, the work criterion, 7 on the second, the criterion of leisure time activities, and 5 on the third, the criterion of mutual confidence. The corresponding figures for Macedonia were: 7, 9, and 8. This was taken to indicate that,
while Matador was quite well teamed for work, Macedonia
appeared to be weakest on this score. On the other hand,
while Matador showed only fair mutuality in leisure time activities, and only relatively little mutual personal confidence,
Macedonia registered a far higher degree of mutual confidence, and an almost optimal amount of mutuality in leisure
time activities.

Since it was initially assumed that our exercise in comparison had a bearing on the problem of cumulative research, it
was necessary to add a brief discussion of this aspect of our
study. Our final argument attempted to show: (1)that cooperative community and society were generically alike and that,
therefore, data gained from the investigation of one had relevance for the study of the other; (2) that the methods employed were scientifically valid beyond the specific subject of
our own exploration; (3) that by accepting as the referential
unit of analysis the construct of the interpersonal relation these
methods made possible cumulative research in sociology.

A final word yet is indicated. As cited by Burgess, techniques available and utilized today by sociological research
are: (1) statistics, (2) personal document and case study, (3)
typology, (4) sociometry, and (5) interviewing.[42] Of these, as
will be noted, we have made use of three: statistics, sociometry, and interviewing—in its modified form of group interview.

Our findings thus remain limited to one of the two basic
aspects of the cooperative community, the objectively given
group structure. But even here we left unemployed typology

[42] Ernest W. Burgess, "Sociological Research Methods," *American
Journal of Sociology,* (May, 1945), p. 476.

and sampling. These two techniques are indispensible where, as for instance, in Israel or Mexico, there exist a great variety of cooperative communities. We could afford to neglect these techniques only because Matador as well as Macedonia, happened to be the only groups of their kind in the given setting.

But the final goal of all research in cooperation is the scientific understanding of what makes people cooperate. Insight related to this kind of understanding can be gained only from the study of the "cooperative personality" structure, the second basic aspect of cooperative community. Here, the personal document and the case study are the given techniques of exploration. Neglect of these techniques—because of limits in time and means—leaves our job half done, and signifies that our labor so far remains preliminary.

TESTING OF A PIONEER TRAINING FARM

The Zionist Pioneer Training Farms

Of all the modern cooperative communities the *kvutza*, or the *kibbutz*, is the most important single source of information for a sociology of cooperation, because of its following characteristics:

(1) Cooperation in the *kvutza* is completely spontaneous in its origin. Membership has always been voluntary in the fullest sense of the term. (2) The *kvutza* practises the highest degree of comprehensive cooperation. Cooperation being all-inclusive, it determines the laws and norms, as well as all economic and social activities. (3) The *kvutza's* bent of mind is experimental. Historically, it is the first modern cooperative community. Its development proceeded in terms of a sociological experiment, conducted by men and women imbued with a spirit of methodical exploration.

The *kvutza* happens to be also the only modern cooperative community which has developed a system of practical preparation for those who intend to join it. The main centers of this preparation are the Zionist Pioneer Training Farms.[1] They were founded, after the First World War, in several countries of Europe and, more recently, also in the U.S.A. and Canada. These farms are essentially replicas of the *kvutza*. They differ from it mainly in that admission to them is less strictly selective, and that membership is limited to the period deemed necessary for effective training in cooperative work and living. (This period varies considerably, depending on conditions. It extends normally to one year. When

[1] See the section on the Training Farm in "Effective Education" in this volume.

during the war, migration to Palestine became difficult, the trainees had to "stick around" for much longer.)

The Aims of Research in Cooperation

The existence of the Pioneer Training Farms makes it possible to investigate the process of cooperation systematically, from its inception. What are the aims of such investigation?

In this instance, sociological research will aim, first and above all at the scientific comprehension of cooperative group structure and its dynamics. As regards the *kvutza* proper, this has been attempted elsewhere.[2] Our present concern is with the specific problems of education for cooperation, as evidenced in the Training Farms.

The most general of the questions to which we seek an answer are: Can behavior be deliberately changed from competitive to cooperative behavior? If so, by what clearly defined means? Is capacity for cooperation essentially uniform, or does it significantly vary from case to case? If capacity varies, can its degree be gauged, and how? What are the criteria by which success and failure of education for a cooperation can be objectively evaluated?

The answers should not only be suggestive generally of how to induce people to cooperate, but of specific value also in the tasks which the Training Farms have set themselves. Are their methods of preparation adequate, or do they stand in need of improvement? Are the criteria by which applicants are selected valid or not? How can it be ascertained whether and to what degree those who are "graduated" from the Farms succeeded better in the *kvutza* than those who never had the benefit of such training?

As we pointed out in our study of the Saskatchewan coop-

[2] Cf. H. F. Infield, *Cooperative Living in Palestine* (New York: Henry Koosis & Co., 1948).

erative farms,[3] our research on cooperation starts with certain basic assumptions. They may be summarized as follows: Between certain essential needs and their satisfaction, our competitive culture has erected barriers which to many individuals seem insurmountable. Some resign themselves to the situation. Others decide to give up competition, with its emphasis on doing things for oneself in rivalry with others, by resorting to cooperation. In so doing, they are ready to yield voluntarily part of their freedom of action in order to share equitably the benefits accruing from their common enterprise.

The key terms in this assumption are "needs" and "barriers". They play a role in the origin as well as in the whole course of the cooperative movement. To the Rochdale weavers, food constituted the most immediate need. They found themselves barred from satisfying that need by the employer on the one side, and by the grocer on the other. The employer paid wages insufficient to cover even their unbelievably modest requirements; and the grocer, to whom they were in perpetual debt, overcharged them cruelly. To escape the dilemma, the Rochdale Pioneers, in 1844, organized what was to become the first successful consumers' cooperative store.

Seemingly insurmountable barriers and powerful needs were present also in the origin of the *kvutza*. They can be traced on two levels: in the immediate situation out of which the *kvutza* grew, and previous to it, in the Diaspora. Wherever he lived in Eastern Europe, political, economic, and social impediments barred the Jew from the normal satisfaction of his needs. Zionism offered the way out by a return to the Homeland, to political, economic and social independence. Once in Palestine, the Zionist pioneers realized that their dominant need was a sound economic basis for their existence. It could

[3] See "Saskatchewan" in this volume.

be found only in agriculture. Again, two barriers blocked the way to the satisfaction of this need: the poor condition of the land, and their own lack of agricultural skills. The barriers were mastered by way of a novel kind of socio-economic organization, the modern cooperative community, the *kvutza.*

The Training Farm forms a part of the well-established *kvutza* system, namely the preparatory stage. It may be assumed that those joining it define their situation in a way analogous to that of the *kvutza* pioneers. As Jews, they experience their life outside Israel as lacking in satisfations. As urban youths willing to do their share in the upbuilding of Israel's agriculture, they are aware of their handicaps. They accept the comprehensive cooperation of the *kvutza* as the most effective way to overcome them. Needs and barriers are, therefore, crucial factors in the scientific analysis of the Training Farms.

The Battery

In the actual conduct of investigation, we employ a "battery" of research devices partly developed and partly adapted by this author in his study of the sociological aspects of cooperative groups. Two of these devices—the test of needs, or what might be called the cooperative potential, and the "obstacles test"— are specifically designed to deal with factors touched upon above. The other techniques: the personal questionnaire, the biographical group interview, and the sociometric test, deal more generally with the same factors.

Let us briefly indicate the nature of these techniques as they relate to the purpose at hand.

A. The Sociometric Test

The general merits and limitations of the sociometric test are discussed elsewhere in this volume. It suffices, therefore, to consider briefly some technical aspects of its application. This test was developed by J. L. Moreno and Helen H. Jennings.[4] It consists of a series of questions—precisely eight in the case of cooperative communities—bearing on the main activities of a given group. A "warming-up" process preceding the actual administration of the test makes it a form of group action. Each member is asked to indicate four other persons whom he would choose for given activities. The choices are recorded in a simple cross-tabulation, both as to number of choices received by each member in the group as well as to choices reciprocated. The number of choices received indicates the "raw sociometric status" of a member, the number of mutual choices is a measure of the degree of sociometric integration. The structure of the group becomes discernible when all the "raw scores" are ordered. This is done usually according to accepted standardization for groups of ten to fifty. The "raw scores" are scaled into five brackets: "stars" (more than 18 choices); "upper" (between 12 and 18 choices); "middle" (between 5-12 choices); "neglectees" (less than 5 choices); and "isolates" (no choice at all). Clearly, an overloading of the extreme brackets (the "stars" on the one hand, the neglectees" and the "isolates" on the other) indicates a poorly balanced structure. Concentration of the majority of scores in the middle sectors, however, may be taken as a sign of balance. A separate recording of the mutual choice scores will give the findings added plasticity and lend itself to graphic presentation in a so-called sociogram.

B. The Cooperative Potential Test

While the sociometric test has proved a reliable instrument for the diagnosis of group structure, particularly in experimental groups which are usually small, it fails to explain the

[4] J. L. Moreno: *Who Shall Survive?* New York: Beacon House, 1934). H. H. Jennings: *Leadership and Isolation* (New York: Longmans, Green & Co., 1943).

divergence in the degree of coherence achieved by different groups. Such divergence is especially puzzling in cooperative communities which, by definition, pursue the same goals and which often are formed in very similar cicumstances by people much alike in background and attitudes. On the assumption that even people desirous of cooperating differ in their capacity for doing so, a test was devised to measure their "cooperative potential."

The test proceeds from a sociological classification of needs into those that can best be satisfied by doing things together with others, and those that can best be satisfiied by doing things for oneself, alone, in rivalry with and competition against others. It consists of twenty-five questions, organized into the five conventional groupings of material, intellectual, spiritual, emotional, and social needs. The testees are asked to rate them in the order of urgency they would assign to them. For example, in the first group the testee is asked to indicate in what order he would want to satisfy the following material needs: (a) the need for tasty food; (b) the need to be able to support those who need his help; (c) the need for proper clothing; (d) the need for adequate housing in a friendly neighborhood. With the ratings scored according to simple statistics and, stated in percentages, we obtain the cooperative potential of a given individual. Summed up and divided by he number of people in the group, the figures yield the potential of the group itself. What the test measures, it must be emphasized, is cooperative *potential*, not actual performance in the cooperative situation.

It has been found useful to arrange the scores of the cooperative potential test in the fashion followed in regard to the sociometric raw status scores. The corresponding brackets are as follows: "extremely cooperative"—72.5 and above; "definitely cooperative"—less than 72.5 but more than 60.0; "fairly cooperative"—less than 60.0 but more than 47.5; "poorly cooperative"—less than 47.5 but more than 35.0;

and "non-cooperative"—less than 35.0. A comparison of the two sets of scores offers clues to the understanding of the differences in group integration. For instance, a group showing low sociometric integration but high cooperative potential, apparently suffers from organizational defects preventing it from actualizing that potential. Take another example. A group shows a cooperative potential which is generally low, but high among those occupying the leading sociometric positions. Relatively good sociometric integration would then indicate, as was found in some cases of "transformations" of capitalistic enterprises into *communities of work,* that those decidedly cooperative are in the lead and able to carry the rest with them. On the other hand, a group with low sociometric integration and equally low cooperative potential is not likely to endure long as a cooperative group.

C. The Obstacles Test

Related to the above test is the test of capacity for overcoming barriers impeding the satisfaction of needs, the obstacles test. In the opening part of the test the respondent is asked to rate his childhood, adolescence and present situation in terms ranging from "extremely happy" to "extremely unhappy". The main part asks the testees to name things he wanted badly in his childhood, adolescence and present situation and how he went about obtaining them. These statements are recorded in terms of passivity, aggressiveness, and indecision, and can be used to estimate the capacity of the group for achieving the cooperative goal. The part of the test related to the present situation serves also to indicate the nature and degree of dissatisfaction existing in the group.

D. The Biographical Group Interview

This is a device designed to obtain a maximum of biographical information about the members and to help them at the

ame time to learn a little more about each other. The inter-
st of the investigator is chiefly in data offered about crisis
xperiences of the members and the manner in which they
ucceeded in overcoming them. Since the cooperative group
epresents a radical departure from competitive culture, ca-
acity for coping with critical situations is very likely to be
significant factor in the success of such a group.

E. The Personal Questionnaire

The biographical group interview, the telling of one's life-
istory in the presence of the group, is usually a highly emo-
ional and integrating experience. Unfortunately, it is very
me consuming and, therefore, not always feasible. In such
ases, the Personal Questionnaire, organized along the pat-
ern derived from the group interview, may serve as the clos-
st possible substitute.

The Setting of the Study

To illustrate the usefulness of the above techniques, we
hall briefly present the findings obtained in the study of one
f the Zionist Pioneer Training Farms in the U.S.A. At the
ime of the study, in 1950, the farm, situated not far from
New York City on 140 acres of flat land, was still in unre-
tricted operation. It practiced mixed farming, milk and eggs
eing its main products. Thirty-five youth of both sexes were
n training on the farm at the time of the study. The two test
essions were preceded by a group discussion in which care
vas taken to explain the nature of the "battery" and to point
ut the significance of the expected findings for the group
tself as well as for the sociology of cooperation.

The prolonged discussion on the first evening left time
only for the Personal Questionnaire. The second evening was
devoted entirely to the three other tests. No time was left

TABLE I
DATA OBTAINED FROM THE QUESTIONNAIRE

	Boys		Girls	
1. Origin				
American Born	B,D,K,N,O,S,U,X,AA,AB,AD,AC,AI	13	A,C,E,F,G,H,J,L,M,P,R,T,V,W,Y,Z	16
Foreign Born	I	1		
		14		16
2. Origin of Parents				
Both Parents Foreign Born	B,D,I,N,O,S,U,X,AA,AB,AC,AD	12	A,C,E,F,G,H,J,L,M,P,R,T,V,W,Y,Z	15
Father foreign born, Mother American	AI	1		
Mother Foreign born, Father American	K	1		
Both Parents American Born			Z	1
		14		16
3. Age				
Between 17 and 18	AI	1	L,V	2
Between 18 and 19	S	1		
Between 19 and 20	U,AB,AD	3	J,T,Y,Z	4
Between 20 and 21	B,O,AA	3	A,C,E,H,P,R,W	7
Between 21 and 22	D,N,X,AC	4	F,M	2
Between 22 and 23	I,K	2	G	1
		14		16
4. Marital Status				
Married	B,D,AA,AD	4	C,E,M,Z	4
Unmarried	I,K,N,O,S,U,X,AB,AC,AI	10	A,F,G,H,J,L,P,R,T,V,W,Y	12
		14		16
5. Education				
High School	I,K,N,S,AA,AB,AD,AI	8	A,J,M,R,V,Y,Z	7
Between 1 and 2 years of College	B,X	2	E,F,P,W	4
Between 2 and 3 years of College	U,O,AC	3	C,G,H,L,T	5
4 years of College	D	1		
		14		16

6. Size of Family

No Siblings	B,I,S,U,X	5	M,T	2
1 Sibling	D,O,AB,AC,AD,AI	6	A,G,H,J,R,V,W	7
2 Siblings	K,N,AA	3	C,E,L,P	4
3 Siblings			F,Y,Z	3
		14		16

7. Family Integration

Well Integrated	B,I,K,O,S,U,X,AA,AC,AD,AI	11	A,C,H,J,L,R,T,W,Y	9
Fairly Well Intergrated	D,AB	2		
Not Well Integrated	N	1	E,F,G,M,P,V,Z	7
		14		16

8. Religious Background

Very Observant	D,I,O,AA,AD,AI	6	E,F,M	3
Moderately Observant	K,U,X,AC	4	A,G,J,R,V,W,Y,Z	8
Non-Observant	S	1	H,L,P,T	4
No Answer	B,N,AB	3	C	1
		14		16

9. Father's Occupation

Businessman or storekeeper	B,K,U,X,	4	C,J,E,G,L,T,W,Y,Z	9
Artisan or Worker	O,S,AA,AB,AC,AD,AI	7	A,F,H,M,P,R	6
Rabbi	D	1		
Farmer	I	1		
Retired			V	1
No Answer	N	1		
		14		16

10. Mother's Occupation

Housewife	I,O,U,X,AA,AB,AC,AD	8	A,C,E,F,H,J,L,P,R,V,W,Y,Z	13
Working Outside Home	B,D,K,S,AI	5	M,T	2
No Answer	N	1	G	1
		14		16

11. Attitudes of Parents towards Pioneer Training

Both Parents Approve	B,D,N,S,U,AA,AD	8
Both Parents Indifferent		2
Both Parents Disapprove	I,K,O,X,AB	5
One Approves, the other Disapproves	AC	1
No Answer	AI	1
		16

	C,F,H,J,L,T,W,Y	7
	P,R	2
	A,E,G,M,Z	5
	V	1
		1
		14

12. Relations with non-Jews

Having non-Jewish Friends of Both Sexes.	B,D,I,K,V,O,S,U,X,AB,AD AI	12
Having non-Jewish Male Friends only	AC	1
Not Having any non-Jewish Friends		.
No Answer	AA	1
		14

	A,C,E,F,G,L,M,P,R,T,V,W,Z	13
	Y	1
	H,J	2
		.
		16

13. Occupational Changes

1 Job	O,S,AB,AI	4
2 Jobs	B,D,K,N,AA,AC	6
3 Jobs	I,U,AD	3
4 and more Jobs	X	1
		14

	A,H,R,V	4
	C,F,L,M,Y	5
	E,J,P,T,W,Z	6
	G	1
		16

14. Length of Time on the Farm

Up to 3 Months	AI,N	2
Between 3 and 6 Months	B,D,I,K,O,AC	6
Between 6 and 9 Months	S,U,X,AA	4
Between 9 and 12 Months	AB,AD	2
No Answer		.
		14

	A,H,L,V	4
	C,J,M,E,R,Y	6
	P,F,T,Z	4
	W	1
	G	1
		16

15. Belonging to a "Kernel"

Belonging to Kernel	I,N,S,U,X,AA,AB	7
Belonging to Kernel Maoz	D,K	2
Not Belonging to any Kernel	B,O,AC,AI,AD	5
		14

	A,H,J,L,R,V,Y,Z	8
	E	1
	C,F,G,M,P,T,W	7
		16

for the Biographical Group Interview. The results produced by the Personal Questionnaire and the tests were as follows:[5]

(a) *The Personal Questionnaire*. The questionnaire was returned by 30 of the members, 16 girls and 14 boys. The results obtained together with the main questions asked are presented in Table I.

Of particular interest was the section on job-changes. It turned out that most of the boys and girls had changed their jobs more than once before entering the farm. The motives for the changes appeared to have no direct bearing on the decision to adopt the cooperative way of living. Although a majority of the respondents declared themselves determined to complete their training and to join eventually a communal settlement in Israel, several expressed doubts as to their fitness for such a way of life.

(b) *The Sociometric Test*. The number of those who participated in the sociometric test was 28, but their choices extended to altogether 30 of the farm's residents. The questions, or "criteria", of the test were essentially the same as those used in other cooperative communities.[6] However, since as trainees the members were rather assigned to work details than given a choice of work, the related criteria had to be modified accordingly. Thus, instead of asking: "Whom do you choose for work?", the question had to be related to the fact that work assignment was handled by a committee to which each of the members had a chance to be elected. It was worded, therefore, as follows: "Whom would you choose to serve with if you were elected to the Work Committee?".

The results of the test are presented in Table II, but the raw social status scores can be ordered as follows:

---

[5] The tabulation of the results was carried out by Mrs. Liuba Syman-Uveeler who assisted in the study and prepared a report published in *Cooperative Living* (Fall 1950).

[6] See "Quantitative Group Comparison" in this volume.

SOCIOMETRIC CHOICE BY CRITERIA

TABLE II

	A	B	C	D	E	F	G	H	I	J	K	L	M	N	O	P	R	S	T	U	V	W	X	Y	Z	AA	AB	AC	AD	BB
A		111	411	040					400					003	200	030				300	200		030		300	100				
B					040			300							120		040			400										
C		400		011		020									200	012	030			100							300	100		
D	030				011															040										
E	033								300						200					130										
F	020		003				004					200		400	200										004		020	400		040
G																														
H		300							300					200	100					100				010				430		
I								300							200			031	010			023		303	102	200				
J									200																					
K			040											400	040						004	024				300		100	400	
L	003					200																						100	211	
M				002										400	200													400		
N																							300							
O	200	200			002																							300		
P							002			034				030	120		021						242	321	030	400		043		
R		404														021				103		003	112			300		400	400	200
S																011	040			200		003	300			322		230	400	102
T									300	022				044	300										003	100		400		103
U																					420							200		
V												300								420								324		
W		300					032			043					100	011										100				
X									400						400															
Y										044					100								200							
Z		400						200						300	122	040	004									111				
AA																		033							211					
AB																						011						140		
AC	043			002																									201	

	A	B	C	D	E	F	G	H	I	J	K	L	M	N	O	P	R	S	T	U	V	W	X	Y	Z	AA	AB	AC	AD	BB
Totals	043													120	120					330					211	402		201		
Work Committee	....	6	3	....	....	....	....	1	9	....	....	2	....	5	16	....	....	....	....	10	1	....	5	2	3	10	1	14	4	3
Walking	3	2	2	1	2	1	1	....	1	6	....	....	....	2	5	6	3	2	5	4	....	3	3	2	2	2	1	5	2	—
Personal Talks	2	2	3	2	3	....	3	....	....	7	....	....	....	2	1	4	2	1	3	1	1	5	2	2	4	3	....	2	2	2
Social Status Raw Score	5	10	8	3	5	1	4	1	10	13	0	2	0	9	22	10	5	3	8	15	2	8	10	6	9	15	2	21	8	5

Number of Mutual Choices: 36

"Stars"	2
"Upper bracket"	3
"Middle bracket"	11
"Neglectees"	12
"Isolates"	2
	30

A majority of the scores thus lies in the extreme brackets. There are 2 "stars", 12 "neglectees" and 2 "isolates". Such distribution of the scores is usually indicative of poor group cohesion. At the same time, the number of mutual choices was 36, which is an extremely high mutuality score. It is worth noting that neither former close affiliation with the Zionist movement nor the length of time spent on the farm appeared to affect the sociometric status of a given member significantly.

(c) *The Cooperative Potential Test.* The five kinds of needs forming the first part of the test were rated by 26 members as follows:

	Boys Choice						Girls Choice				
	1st	2nd	3rd	4th	5th	No Answer	1st	2nd	3rd	4th	5th
Material Needs	3	–	6	–	1	1	1	1	–	8	5
Intellectual Needs	2	4	1	2	–	2	2	6	5	2	–
Emotional Needs	2	1	2	1	3	2	7	3	5	–	–
Spiritual Needs	1	–	–	3	5	2	–	–	2	3	10
Social Needs	2	5	–	3	–	1	5	5	3	2	–

In contrast to the boys, the girls tended to rate material needs lower than emotional ones. Spiritual needs received the lowest ratings from both sexes.

The cooperative potential scores ranged as follows:

	Boys		Girls	
Extremely Cooperative	I, K, AC, B......	4	Y, H, E, C......	4
Definitely Cooperative	AA, O...........	2	V, M, T........	3
Fairly Cooperative	AB, U, D........	3	Z, L, W, F, P, J	6
Poorly Cooperative	S, N............	2	R, A...........	2
		11		15

The group's total cooperative potential score was 61.4%

(d) *The Obstacles Test.* The number of forms returned on this test was 21. Of particular interest to us were the ratings in the first part of the questionnaire and the statements offered on what the members "wanted badly" in their present situation.

The question asked in the first part of the test was: "Would you say that you have been: thoroughly happy, mostly happy, mostly unhappy, thoroughly unhappy (a) in your childhood, (b) in your adolescence, (c) since on the farm?" The answers were as follows:

	In your childhood	In your adolescence	Since on Farm
Thoroughly happy	5	4	2
Mostly happy	14	11	17
Mostly unhappy	1	6	2
Thoroughly unhappy	1	0	0

Only seventeen of the members cared to answer the last question. Of these, eleven stated that what they "wanted badly" since on the farm was "to be accepted by the group". One other member wanted "to see more of a social entity". Still another desired "to see some of the group display less selfishness and more sincere regard for the group as a whole". Complaint was registered also that "people here compete with each other ruthlessly", or that "people who represent selfishness and falseness in character . . . seem to be stronger than those who are basically 'good.' "

Interpretation

The above results were obtained in one of several studies conducted in the Zionist Pioneer Farms in the U. S. A. The studies demonstrated to those responsible for running the Farms the value of the "battery" as a fact-finding device. It was able to secure, in the shortest time possible, a considerable number of facts useful to the administration of the Farms. That demonstration being the immediate purpose of the studies, the discussion of the results with the group themselves, wherever held at all, was incidental to that purpose.

An examination of the results reveals certain debilities in the work of the Pioneer Farms generally as well as in the specific case cited above.

As for the more general weaknesses, the results of the Personal Questionnaire suggest, among other things, that the appeal of the Pioneer Organization was limited to a marginal section of the population concerned. Most of the 30 respondents, or 27, were children of foreign born parents. Two were of parents of whom one was foreign born; one of the respondents was foreign-born himself. If these findings can be taken as representative—and they were confirmed by data obtained from other Farms—they reveal a critical narrowness in the range of the organization's appeal. It may be assumed that the leaders of the organization were not unaware of this shortcoming. The significance of the results obtained by means of the Questionnaire lies in the fact that they demonstrate the peculiar demographic character of the youths willing to join the Farms. In this situation, those responsible for formulating the recruiting policy of the organization may decide to do either of two things: resign themselves to having the farms attract the present type of youths; or reconsider the ways and means of extending their appeal to the norm of the American Jewish population.

More specifically, the tests helped disclose a serious discrepancy between the avowed aspirations and the actual achievements of the group with respect to its cooperative intentions. At first sight, the results of the sociometric test appeared to be contradictory. On the one hand, the "raw social status" scores indicated poor group cohesion. On the other hand, the score of mutual choices, the "mutuality score", turned out to be extremely high. Actually, the two scores reinforced each other. The poor group cohesion was found to be due mainly to excessive clique formation. The high mutuality of choices in these sub-groups lowered the cohesiveness of the group as a whole. But a real contradiction appeared when the results of the Sociometric Test were compared with those of the Cooperative Potential Test. While the former indicated low group cohesion, the latter showed that the potential of the group for cooperation was decidedly strong. In other words, those who formed the group wanted and were fairly well capable of cooperation; but their needs and capacities found no opportunity for satisfaction in the present situation. Obviously, something must be basically wrong with a group which, deliberately organized for cooperation, remains so far from realizing its cooperative potential.

In a case like this further diagnostic clarity may be derived from a closer scrutiny of the test results. It might be found that the cause of frustration lies with the leadership. The scores may reveal that the extremely high number of choices the "stars" receive, was not evenly divided among the different criteria. If, as in the present case, the number of choices on the work criterion by far outweighs those on the criterion of personal mutuality (Cf. our Table II under O and AC with a distribution of 16 to 5 and 1, and 14 to 5 and 2 respectively—) and if, at the same time, the cooperative potential score of a given "star" is significantly low, it is fair to conclude that the leaders are either incapable or unwilling to foster genuinely cooperative relations within the group. In the present case, this conclusion was also borne out by the statements offered in the Obstacles Test which contained almost wholly complaints about the lack of cooperative spirit in the group.

It appears that a skillful use of the battery can be helpful not only in securing a wealth of pertinent data, but also in locating existing debilities in the cooperative group structure. Diagnostic findings of this kind may, in turn, form the basis for effective remedial group action. That was systematically attempted in the study of the French *communities of work* that follows.

13

EXPERIMENTAL GROUPS AND SOCIOLOGICAL COUNSELLING

•

The Need of Experimental Groups for Counsel

To be asked for counsel and to find his services as consultant sought after is for the sociologist something of a novel experience. The demand, prompted by the exigencies of World War II, and restricted mainly to government agencies and large-scale business, may be small but it is definitely growing.[1] If the trend continues, it may produce a significant change in the professional standing of sociologists. Instead of exclusive devotees to research and education, as in the past, they may become practitioners much like consultant engineers and consultant psychologists. Those disturbed about the academic inbreeding of sociology will undoubtedly welcome the change. Others will foresee certain dangers. Chief among them will be the risks to which any science exposes itself when it undertakes to serve narrow and non-scientific concerns. It would seem that sociology might avoid such risks if it applied itself, first and primarily, to groups and people whose concern was similar to that of science, such as experimental groups. Most important among these are the modern cooperative communities, such as the Israeli *Kibbutzim,* the Saskatchewan cooperative farms, and the French *communities of work.*

The opportunities these groups offer for sociological inquiry have been commented on before. Harder to recognize is the need of such groups for the kind of assistance sociology is able to offer. For those groups, proceeding more or less in line with the canon of sociological experiment, not only initiate the experiment but serve, at the same time, as the very sub-

[1] See Wellman J. Warner, "The Role of the Sociologist." *Bulletin of the American Sociological Society* (Sept. 1951), particularly Table VII, p. 9.

stance of its conduct.[2] Personal involvement, which is, on the one hand, indispensable for the conduct of the experiment, is, on the other hand, virtually incompatible with the requirement of scientific detachment. Experimental operation is only one, though the most essential, part of experiment. Its full value will hinge upon attention to other requirements of the experimental method, such as formulation of a consistent theory stating the initial assumptions explicitly or implicitly; close observation and faithful recording of each step in the progress of the experiment; valid interpretation of the results obtained; and all the logical, statistical or other mathematical manipulations which lead to predictive generalizations forming the ultimate goal of all experiment. It is hardly likely even for the best qualified members of such groups to satisfy these requirements themselves. The trained expert will have to be called upon to supply them.

For the sociologist, the main difficulty in such a situation lies in the fact that, in order to lend the required assistance, he stands in need of skills and tools with which his discipline has only poorly equipped him. Since he will have to know not only how to handle groups but how to treat them as well, the capacities at his command will have to range from ability to assume leadership to the skill of making painful truth palatable. No formal academic training is likely to have taught him these aptitudes. To develop them, he will have to fall back upon his own ingenuity and resourcefulness. An added difficulty results from the fact that at present cooperative communities are located far apart from one another. Their own funds as well as those available for research of this type are not ample. To make the best of available time and means the sociologist will need instruments that will help him obtain reasonably precise data without prolonged application. Again, it will be mostly up to him to devise such tools.

Having their attention drawn to a set of tests used in the

[2] Cf. "Utopia and Experiment" in this volume.

GFRI's cooperative community research[3] in an article written for their own magazine, *Communauté,* by Claire Huchet Bishop,[4] the leaders of the *Entente Communautaire,* the federation of the French *communities of work,* invited this writer to apply them to their own groups. Subsequently, a study of the communities was carried out, with the invaluable assistance of Mrs. Bishop. In all, eight communities, selected so as to form a sample representative of all existing communities, were visited in the spring of 1951. The following description of the work done with one of them, to which we shall refer by the fictitious name of *Clermont,* will serve to indicate the nature of the study as a whole.[5]

The Setting of the Study

Clermont is one of the most important of the French *communities of work.* It is one of the largest, most successful, and most fully developed among them. The organizational and administrative form of a *community of work* has been described in detail elsewhere.[6] It might be helpful, though, to recall briefly the features relevant to the discussion that follows.

Though a cooperative community, *Clermont* is not a community in the same sense as a *Kibbutz.* The men and women belonging to *Clermont* do not form a settlement apart from other people. They live dispersed throughout a city of some 30,000 inhabitants, in the South East of France. Their community identifies itself in terms of a specific social organization based on a common economic enterprise, the production of certain metal parts. The factory is located in the outskirts of the city. It is equipped with the most modern machinery

[3] Cf. "The Testing of a Training Farm," in this volume. Also Liuba Syman-Uveeler's "Report on the Study of a Pioneer Training Farm," *Cooperative Living,* II, 1.

[4] "Les Communautaires Américains," *Communauté* (January 1951).

[5] A full report on this study is in preparation under the tentative title of *Sociological Counselling in Experimental Groups.*

[6] See "The Urban Cooperative Community," in this volume.

—some of it devised by *Clermont's* own enginering staff. It is housed in well-lighted and sanitary buildings. Production is up to the highest standards attained in that branch of French industry and accounts for a fifth of the country's output. The specific character of *Clermont* manifests itself, however, in the non-technical facilities that are part of the factory. Main among them are: the assembly hall, which serves also for social and educational activities; the non-profit canteen where all who wish may take their meals; the offices of the Social Section; the library; and the nursery, which accepts children of non-members, too.

Membership in the community is of two kinds: The *productifs*, or productive companions, men and women who work in the factory or are in any capacity connected with the affairs of the community; and the *familiers*, or family companions, those married to a productive companion. All the family companions who took part in the tests were women. To become a companion, a man or woman must pass through the two consecutive probationary stages of apprenticeship and candidacy, the first lasting three, and the second thirteen months.

The most characteristic unit of *Clermont's* social organization is the *groupe de quartier*, or Neighborhod Group. This unit is not a part of the administrative framework. Its main function is social. It serves to create the group coherence among the ecologically dispersed membership that is an indispensable attribute of a community. If successful, it fosters personal intimacy and turns cooperation from a mere way of doing business into a way of life. So far as the communitarian aspects of *Clermont* are concerned, the neighborhood group is of crucial significance. It has rightly been called the community's elemental cell.

The Tests Applied

Although it may not be a strict requirement for sociological testing that it be applied simultaneously to all members of a group, it is certainly of great advantage to do so. It saves time and energy; it helps obtain the fullest data; and, by lending the procedure the character of group action, it tends to make the results, particularly in a communitarian group, more significant. Where, as in the case of *Clermont,* it proves for technical reasons too difficult to handle the group as a whole, it is necessary to apply the tests to smaller units. The division into units should follow, if possible, some already existing lines of demarcation. The basic distinction between companions by virtue of work contribution, on the one hand, and by virtue of marriage, on the other, offered itself here as a natural dividing line. Consequently, the tests were given in four sessions, two with productive, and two with family companions. According to the official lists, the full adult population of *Clermont* consisted of 196 persons. Of these, as may be seen from the following Table, 100 men and 15 women were productive, and 63 were family companions.

TABLE

	Male	Female	Total
Productive companions	100	15	115
Candidates	5	0	5
Apprentices	11	2	13
Family companions	0	63	63
Total	116	80	196

The leadership of the community decided that companions only should participate in the tests. Three candidates, however, requested and were granted special permission to take part. The sessions took place in the assembly hall where the participants were seated on benches along wooden tables. The order of application was identical with that followed in the other communities. The first session in each of the two sub-groups was devoted to the Sociometric and the Cooperative Potential Test; the second, to the Obstacles Test and the

Personal Questionnaire. Each session was opened with a few introductory remarks designed to explain the purpose of the study in general. When the forms were distributed, the participants were asked to first acquaint themselves with the content. All printed matter was read aloud to the group. After the reading, the participants were encouraged to ask questions and to make sure that the content of the tests was completely clear to them.

Personal Background

The Personal Questionnaire, which is actually a substitute for the Biographical Group Interview, is used to secure, within a minimum of time, data on the personal background of members in cooperative communities. Response to it was inhibited in all communities studied, mainly because of the strong suspicion with which French workers, several years after the German occupation, still regard any questioning of this sort. The answers received were, therefore, far from complete. They are suggestive rather than an accurate picture of the members' personal background. The data may be summed up separately for each of the two sub-groups, as follows:

A. *The productive companions.* Of the 73 productives, 60 men and 13 women, who answered the question about their origin, all were French born of French parents, with the exception of three who were French born of non-French parents. Of the latter, two were of Italian and one of Armenian descent. The dominant age level of the productives was between 25 and 35 years of age. More specifically, 27 of the respondents, 22 men and 5 women, were between the ages of 25 and 30, while 23 (20 men and 3 women) were between 30 and 35. There were 6 (4 men and 2 women) who were under 25; 9 men were between 35 and 45, two men and one woman between 45 and 50; and one man and one woman between 50 and 60. 55 men and 3 women were married, 4 men and 6 women were not. 3 were widows. 23 men and 7 women had only primary education; 11 men and 2 women had finished the French equivalent of high school; and 14 men, but none of the women, had attended a technical high-school. Only 9 men and no women were graduates of an institution of higher learning.

As to size of family, fourteen men and three women came from one-child families. As to siblings, the numbers were as follows: 1 sibling: sixteen men and three women; 2 siblings: nine men and one woman; 3 or 4 siblings: seven men; 5 siblings: three men and two women; 6 siblings: two men and one woman; 7 or more siblings: two men. Only 26 men and 4 women answered the question regarding family integration. Of these, 25 men and 3 women considered the families they came from as well integrated, and one man and one woman as not well integrated. The number of those answering the question about the father's occupation was slightly higher, 40, including 35 men and 5 women. 12 of the 35 men stated the father's occupation as manual worker; 6 as farmer or civil servant, respectively; 4 each as businessman or store-keeper; 2 each as office worker; and 1 as artisan. 4 listed the father's occupation as professional, without specifying the profession. Of the 5 women, 2 gave the father's occupation as civil servant; and 1 each as manual worker and farmer. The mother's occupation was given by 25 men and 2 women as housewife. 9 men and 2 women marked it as "working outside the home." Of the 45 answering, all but one man and one woman indicated that they also had friends outside the community.

The fullest data were obtained on occupational changes prior to employment at *Clermont*. This was probably due to the fact that the importance of this point was stressed in the introductory remarks. Of the 58 men and 9 women answering, 29 men and 7 women had not had more than one job; 18 men and one woman reported 2 jobs; two men and one woman 3 jobs; and three men 4 and more jobs. For two men, their

jobs at *Clermont* were the first in their careers. Four men
gave as their previous occupation: work as deportees in Germany during the war.

B. *The family companions.* All of the family companions
answering the questionnaire were French born of French
parents, except two whose parents were Italian and one
whose parents were Spanish. The main age range was, here
too, between 25 and 35, twelve being between 25 and 30, and
seventeen between 30 and 35. None was under 20, and three
between 20 and 25; three between 35 and 40; two between
40 and 45; and one between 50 and 60. All of the 43 respondents were married. 30 had completed primary school,
4 had attended high-school, and 6 a technical high-school.
None listed attendance at an institution of higher learning.
Eight grew up as only children; each of nine had one sibling;
nine listed 2, and five 3 siblings; two came from families with
5, and one from a family with 7 or more siblings. Only 14
of the family companions answered the question about family integration, and all rated it as well integrated. As to the
father's occupation, 9 stated farmer; 8 each office or manual
worker respectively; 7 civil servant; and one each artisan or
professional. 16 gave the mother's occupation as housewife,
and 12 as "working outside the home." None of the family
companions indicated any change of job prior to marriage.
11 had earned their living as manual workers, 8 as seam-
stresses, 7 as typists, 3 as house-maids, and 2 as salesgirls.
One each had worked in civil service, a pharmacy, a beauty
parlor, and as telephone operator.

To sum up: the data obtained through the Personal Questionnaire are by no means as complete as they might be.
But they are sufficient to demonstrate that the membership of
Clermont is, on the whole, not atypical, exhibiting the characteristics of the average French working class population.
The community's population is, so far as our sample goes,
100 per cent French by birth, and working class by family

background, education, and occupational career. The only exceptions, possibly, are the nine men who received higher education and are highly trained technicians. Several of the latter came to the community as hired professionals. Those who decided to stay applied for membership.

The Sociometric Group Structure

A. *The productive companions.* Of the total of 115 productive companions, seventy-one returned a completed sociometric questionnaire..[7] The actual number of participants was larger. Some fifteen questionnaires were incomplete and had to be discarded. On the other hand, several companions who could not attend the session were allowed to fill out the forms individually. The questions asked were—re-translated from the French—as follows: If you had a choice, whom would you choose: (1) to work with on the same team? (2) to take part in your neighborhood group? (3) to sit next to you at meals? (4) to go fishing or hunting? (5) to go for walks? (6) to go to movies or the theatre?[8] (7) to discuss the affairs of the community? (8) to discuss personal problems? In short, the number of questions was the same as that used in the other community studies. Their wording was adapted to the specific circumstances of the *community of work.* According to the standard procedure followed in sociometric tests, the participants were asked to answer each question by placing in the order of preference no more, and, if possible, no less than four names in the space provided for this purpose, and to give first thought to members of the community. If a family member or a person outside the community were chosen, this was to be noted appropriately.

As in previous studies, only three of the questions were

[7] We propose to disregard here the three candidates.
[8] The fourth question was extended verbally to include camping, and the sixth question to include other public entertainment, such as concerts etc.

used for scoring. These were questions 1, 5 and 8, corresponding to the three criteria of "work," "going for walks," and "discussing personal affairs." Tabulation of the results showed that the choices of the seventy-one respondents extended to altogether 140 members of the community or, more specifically, to 115 productive companions, five postulants, thirteen apprentices, and seven family companions. The choices relating to family members—67—and to persons outside the community—26—were not used in the raw social status scores.[9]

Because of limitation of space, we present only the final results, as they appeared in the tabulation. In arranging the results, the standardization established by Bronfenbrenner[10] is followed. Because of the larger number of "neglectees," we distinguished those who received five or less choices from those who received no choice at all, and put the latter into a special bracket of "isolates." Ordered in this way, the raw social status scores of the productive companions shape up as follows:

TABLE II

Bracket	Number
"Stars," or those who received more than 18 choices	3
"Upper," or less than 18 and more than 12 choices	2
"Middle," or less than 12 and more than 5 choices	23
"Neglectees," or less than 5, but at least 1 choice	78
"Isolates," or no choice at all	9
Total	115

The total number of mutual choices was 37.

[9] The total number of available choices was 71 x 12, or 852. Of these 594 went to members of the community, and 93 to others, as stated above. Thus the total number of choices made use of was 687.

[10] See Urie Bronfenbrenner, "A Constant Frame of Reference for Sociometric Research," *Sociometry*, VII (1944), p. 44.

B. *The family companions.* The number of completed questionnaires returned by this sub-group was 51, The choices extented to sixty-three family companions, thirty-three productive companions, and two candidates, or altogether ninety-eight members of the community. As before, the choices that went to family members—59—and to persons outside the community—22—were disregarded.[11]

If ordered in the manner described above, the scores of the family companions present themselves as follows:

TABLE III

Bracket	Number
"Stars"	0
"Upper"	0
"Middle"	4
"Neglectees"	52
"Isolates"	7
Total	63

The total number of mutual choices was 14.

C. *Cross tabulation.* As indicated above, there was a certain amount of cross-reference in the choices between the two sub-groups of the productive and family companions. While only seven family companions received choices from the other sub-group, the number of productive companions was thirty-one. We may use these cross-choices to correct the raw social status scores of those who received them. These corrected scores would indicate sociometric status as it refers not only to the sub-group, but to the community as a whole. As far as the productive companions are concerned, the addition of

[11] The total number of choices available in this instance 51 x 12, or 612. Of these there were utilized 276, distributed as follows: 137 went to family companions; 58 to productive companions and candidates; and 81 to others.

choices received from the family companions alters the bracket of status in seven cases: one moves from the "Upper" to "Star" bracket; three from "Middle" to "Upper"; and three from "Isolates" to "Neglectees."[12] The others "improve" their position, but remain within the same bracket. Thus, the corrected picture for the productive companions is as follows:

TABLE IIIa

Bracket	Number
"Stars"	4
"Upper"	4
"Middle"	20
"Neglectees"	81
"Isolates"	6
Total	115

Hence, in spite of a reduction in the number of "Isolates," the total for "Neglectees" and "Isolates" remains the same.

By far less marked is the improvement on the side of the family companions. Only one woman—who interestingly enough, received only two choices from her own sub-group, but five choices from the productive companions— moves on the basis of this correction out of the "Neglectee" into the "Middle" bracket. There is some improvement in the status within the " Neglectees" bracket, but no change in the number of "Isolates." Thus there is only one change in the overall picture of the family companion scores. There are now five in the "Middle" bracket instead of four; but the number of "Neglectees" and "Isolates" remains unchanged here too.

As to mutual choices between the two sub-groups, there is only one such choice.

[12] It is interesting to note, that in one case, that of a woman productive companion, the rank improvement is considerable. While receiving no choices from the productive companions, she was the recipient of five choices from the family companions.

The Cooperative Potential

A. *The productive companions.* Eighty-three of this sub-group returned cooperative potential test forms fit for scoring. The results were ordered, in correspondence with the five brackets of the sociometric scores, as follows:

TABLE IV

Bracket	Number
I. Potentially extremely cooperative (above 72.5)	6
II. Potentially definitely cooperative (less than 72.5, but more than 60)	14
III. Potentially fairly cooperative (less than 60, but more than 47.5)	26
IV. Potentially poorly cooperative (less than 47.5, but more than 35)	22
V. Potentially non-cooperative (less than 35)	15
Total	83

The extremes ranged from 82.3 to as low as 14.6. The cooperative potential score for the sub-group as a whole was 49.2.

B. *The family companions.* The scores of the forty-seven members of this sub-group who returned forms properly filled out, were as follows:

TABLE V

Bracket	Number
I. Potentially extremely cooperative	1
II. Potentially definitely cooperative	6
III. Potentially fairly cooperative	20
IV. Potentially poorly cooperative	8
V. Potentially non-cooperative	12
Total	47

The highest score was 73.3; the lowest 22. The cooperative potential score for the sub-group as a whole was 47.5.[13]

The Obstacles Test

A. *The productive companions.* We are limiting ourselves to a summary recording of the results of this test. Here are the figures referring to self-rating on happiness in childhood, adolescence, and in the community:

TABLE VI

Rating	Childhood	Adolescence	Community
Very happy	26	12	17
Mostly happy	36	45	50
Mostly unhappy	10	15	5
Extremely unhappy	0	0	0
Total	72	72	72

We see that the number of those who rate themselves as "very" or "mostly happy" in all three instances is: 62, 57, and 67, respectively. The numbers for the "mostly unhappy" rating are: 10, 15, and 5, respectively. Of the five who rate themselves as "mostly unhappy" in the community, one rates himself as such throughout and another in two instances. Of the fifteen who rate themselves "mostly unhappy" in adolescence, all but the same two rate themselves as "mostly happy," and in one instance as "very happy" in the community.

An admittedly loosely defined and rather tentative attempt

[13] It might be interesting to note that the following needs were listed on the last page, left free for the statement of additional needs:

Need	Times Mentioned
Peace	45
Justice	10
"Americans back to America"	10
Brotherhood of all men	6
More time for the family	6
To learn more about Americans who are against war	2

was made to evaluate in terms of resoluteness the rather lengthy and often caustic statements in response to the three main questions of the test, as to obtaining things badly wanted in the three periods of life. The results can be stated as follows:

Rather more resolute:	39
Rather more irresolute:	21
Undetermined:	12
Total	72

If classified according to their self- or community centeredness, the things badly wanted may be grouped as follows:

Number of those who express self-centered wants	31
Number of those who express community-centered wants	23
No wants stated	18
Total	72

The number of community-centered wants stated by the twenty-three productive companions was altogether forty. Listed in the order of frequency with which they were mentioned, they are:

TABLE VII

Nature of Wants	Times Mentioned
More solidarity among the members	4
Increase in the percentage of remuneration	4
Music	4
Assigning more responsibility to members	3
More mutual confidence	3
More tolerance towards others	3
A Farm	2
More sincerity	2

More attention to the needs of the rank and file	2
A full month of vacation	2
More sport	2
Special control for the industrial service	1
Less concern with money	1
Fewer speeches and meetings	1
More restraint on the part of the chief of community	1
A car for outings and transport of children	1
More politeness towards those who speak in meetings	1
More order and good manners	1
More mutual helpfulness	1
Less anti-feminism in the community, particularly on the part of the chief of of community	1
Total	40

B. *The family companions.* The corresponding data for the family companions are as follows:

TABLE VIII

Rating	Childhood	Adolescence	Community
Very happy	12	4	1
Mostly happy	17	23	31
Mostly unhappy	7	9	5
Very unhappy	2	1	0
Total	38	37	37[14]

The number of those who rate themselves as "very" or "mostly happy" in all three columns is: 29, 27, and 32, respectively. The number of those who rate themselves as

[14] The difference in the total is due to omission on the part of two who gave ratings in the two other columns.

"mostly unhappy" is 7, 9, and 5, respectively. There are two who rate themselves as very unhappy in childhood, and one in adolescence, and none in the community. All who rate themselves as "mostly" or "very unhappy" in adolescence, rate themselves as " mostly happy" in the community. Of the five who rate themselves as "mostly unhappy" in the community, one adds a note to his rating saying "ça va mieux," or "there is some improvement."

An evaluation of statements on obtaining things badly wanted, as in the case of the other sub-group, yielded the following figures:

Rather more resolute:	12
Rather more irresolute:	22
Undetermined:	4
Total	38

As to the self- or community-centeredness of the wants:

Number of those who express self-centered wants:	12
Number of those who express community-centered wants:	26
No wants stated:	0
Total	38

Twenty-six specific community-centered wants were named by the twenty-two family companions. In the order of frequency, they may be listed as follows:

TABLE IX

Nature of Wants	Times Mentioned
Genuinely cooperative living based on more mutual confidence	6
A housing policy	6
Less difference in wages	5

More opportunity for education through liberation from household duties	5
Peace	3
More sincerity among the members	1
Total	26

Discussion of the Results

It is well to remember the limited scope of the present study when discussing its results. It did not intend to cover all significant aspects of the enterprise. The immediate reason for the study were certain difficulties encountered by the community. Main among them was that of establishing standards of social evaluation as precise as those for work efficiency. The tests offered themselves as a possible basis for solving this problem. Even though the leaders of the Federation welcomed the offer, their attitude towards the tests was rather one of skepticism. To demonstrate the soundness of the method proposed became, therefore, the first and, it was thought, preliminary objective of the study. It so happened, however, that the tests revealed a disproportionately large amount of stress in the internal social system of the community. This disclosure proved to be startling, and concern with appropriate remedial action so urgent, that the original objective had to be left in abeyance at least for the time being.

A brief consideration of what the test results mean will make their startling effect upon the group understandable.

To begin with the results of the sociometric test, they indicated a strikingly low degree of social integration. Of the 115 productive companions, according to Table IIIa, only 24 were in the "upper" or "middle" bracket, while 4 were "stars" and 87 "neglectees" or "isolates." Even worse was the case of the family companions. Table III shows that there were none in in the "upper" and only 4 in the "middle" bracket, while —

of a total of 63, 59 were "neglectees" or "isolates." Even if the standards are reduced and all those with a score of 5 are moved into the "middle" bracket—a procedure possibly justified by the larger spread of choices and by the marginal position of this score—the picture does not improve substantially. There were 18 productive and 2 family companions showing this score. Thus the number of "neglectees" and "isolates" in both sub-groups remained high enough, 69 and 57 respectively, to mark sociometric integration as extremely poor.

Looked at more closely, the results of the sociometric test become even more ominous. Apart from the number and distribution of the choices, the nature of the choices in itself seems to suggest that the community is moving away from its original goal. It must be remembered that *Clermont* was formed by a group of workers who, by designating their novel undertaking a *community of work* signified that they were putting accent on "community" rather than on "work." Work was supposed to be important only in so far as it was indispensable in securing a material basis for the community. Work was to remain subordinated to the pursuit of the main goal: satisfaction of the industrial worker's need for social reintegration, with all that it implies in the way of "we-feeling," sense of belonging, and personal dignity. A sociogram, drawn in terms of mutual choices in both sub-groups and based on the three criteria of "work," "going for walks," and "discussing personal problems," yielded, however, the following picture: of the altogether 74 mutual choices registered by the productive companions, 48 were based on the criterion of "work," and the rest, 26, on the two other criteria indicative of personal intimacy, or friendship; while on the side of the family companions, all 28 mutual choices were limited to the one criterion of "work." Since the main accent of mutuality was thus on work, it was not surprising to find that there was virtually no mutual choice connecting the two sub-groups. The only mutual choice recorded was one on "going for walks" and was made by a man and a woman who belong to the same neighborhood group.

All in all, the results of the sociometric test showed that the community as a whole was very poorly integrated; that such mutuality as developed was confined mainly to work-relations; that there was little mutuality between the productive and the family companions; and that so far as the latter were concerned, they had not yet reached the point where they could be considered either as forming a group of their own or being integrated in the community as a whole. In short, in the light of the results of the sociometric test, *Clermont* presents itself as a group of people who work well together, but who have hardly begun to achieve that social reintegration for which their community specifically was organized.

For an explanation of the poor state of social integration of *Clermont,* let us look at the scores of the cooperative potential test. Table IV shows that of the 83 productive companions, 43 rank between potentially "fairly" to "extremely" cooperative, with a smaller number, or 37, ranking as potentially "poorly" or "non-cooperative." The corresponding figures for the family companions are, as may be seen from Table V, 27 and 20. If these scores are held against those of the sociometric test, the conclusion suggests itself that the group's potential for cooperation, although it may be qualified as only fair on the whole, is nevertheless superior to the actualization it found within the community.

This conclusion is especially suggestive in the case of the family companions. It has an important bearing on an issue that was found present in all the communities we studied, with the one exception of a rural community. One of the main obstacles encountered by all these groups relates to the integration of the women of the community. The blame is usually placed on the alleged nature of the French woman who is said to be incapable of cooperation. *Clermont* was no exception in this respect. But the findings of the cooperative potential test presented a strong argument against such claims. The total scores for the two sub-groups were about equal. They were: 47.5% for the family, and 49.2% for the productive companions.

It may be fair to assume, then, that there existed at *Clermont* a discrepancy between the cooperative potential and the opportunity it had for realization. An indication as to how this discrepancy impeded the growth of what the members like to call "the communitarian spirit," may be found if some of the more significant individual cases are considered. Productive companions with the highest sociometric scores on the one hand, and those with the highest cooperative potential scores, on the other, may well serve the purpose. There are, as Table IIIa shows, 4 "stars" and 4 in the "upper" bracket. Identified by letters used in the general tabulation and presented with the number of choices received as well as with their cooperative potential scores, they are as follows:

TABLE X

Bracket	Name	No. of choices	Coop. Potential
"Stars"	CM	23	50.6
	J	22	82.0
	AT	22	52.3
	X	21	70.0
"Upper"	AI	14	38.0
	BZ	14	63.3
	CW	13	54.0
	CF	13	58.0

If on the other hand, the sociometric status of an equal number of those who show the highest cooperative potential scores is considered, the figures are as follows:

TABLE XI

Bracket	Name	No. of choices	Coop. Potential
Extremely	AW	2	82.3
cooperative	B	4	82.0
	J	22	82.0
	AV	3	80.0
	BN	1	80.0
	H	3	74.6
Definitely			
cooperative	X	21	70.0
	AN	11	70.0

In other words, in its trend toward economic expansion, the community appeared to stress work capacity over cooperative ability. That affected not only its selection of prospective members, but also influenced its choice of those it put in the lead—a vicious circle, by which strength is turned into anxiety. The results of the obstacles test only confirm the presence of that anxiety. Both Tables VI and VII, indicate that those joining the community experience a definite increase in satisfaction. On the other hand, there is the long list of things wanted badly and which are missed at *Clermont*. Significantly enough, there are some eminently communitarian items on that list, such as solidarity, mutual confidence, tolerance, a more equal sharing of responsibilities, and the classic demand for less "anti-feminism" raised by one of the productive companions. To which there might be added the even more acute, but equally unsatisfied wants of the family companions, such as the repeatedly stressed desire for a "genuinely cooperative living based on mutual confidence" and for "more opportunity for education through liberation from household duties."

Presentation

Sociological tests should serve two purposes at one and the same time. As tools of research, when applied to experimental groups, they must discover, more precisely than would be the case otherwise, certain crucial data about the internal social system of such groups. In this respect, the above tests may serve mainly as fact finding tools. Yet, even when used in that way, the tests cannot be administered effectively without the active and, it may be added, animated participation of the people concerned. By participation of that kind, the tests become instruments of group action.

The people themselves are naturally less concerned with research as such than with the succes of their experiment. They are interested in the tests only in so far as the results may contribute to the achievement of their goal. They will value the tests not so much because of the facts revealed thereby but as instruments enabling them to make a better job of what they are trying to do. Stock-taking is a prerequisite of improvement. In other words, the group will judge the tests by their diagnostic competency. The presentation of results accordingly will assume the meaning of a consultation that, in the case of any weaknesses revealed, might lead to remedial action.

This is what actually happened at *Clermont*. Discussions with the leaders of the group and with the rank and file, particularly at meals in the canteen, confirmed the correctness of that approach. Its dramatic effect will be appreciated better if the rankling opposition against the tests as such and those administering them is remembered. Among the obstacles to overcome, there was, first, the common species known as "test-resistance" which, for whatever reason, is somewhat more acute among Frenchmen than among other nationals studied. Second was the suspicion with which many French workers are presently inclined to view anything American. An active clique of political doctrinaires did its best to add fuel to the antagonism. To complicate things further, certain mistakes were made in handling the group. For instance, the first test session had to be held without a sufficient "warming-up" process, something that may prove to be injurious to all testing of experimental groups. Then again, because of the size of the group, it was found that assistance would be needed in tabulating the results. A request for such assistance from among the members themselves was granted only after some publicly expressed suspicion that the request violated the promised confidential treatment of the results. All during the tabulation, which took more than a week, the group was greatly agitated about the whole study, with factions forming for and against it, and exchanges taking place between some of the men and the head of the community. Thus, the day of presentation arrived amidst a great deal of apprehension.

The presentation was made at the weekly "Assembly of Contact". Usually only the productive companions participate in that Assemby. Since this was a special occasion of concern to the community as a whole, the family companions were also well represented. Altogether, some two hundred persons attended. The procedure adopted for this session was modelled after that advocated by recent research on doctor-patient relations, holding that optimum results are obtained when the patient is given a chance to take an informed part in each step of his own examination. All the tabulated material was displayed and made easily accessible to the audience. Reference was made to that material throughout the presentation, and care was taken to elucidate the manner in which the results were obtained and interpreted. The gist of the presentation may be summarized as follows:

First, a few introductory words were said about the nature of the tests and the work of the GFRI in general. It was pointed out that the tests served but as tools, similar to the

instruments used by a physician, and that their usefulness
depended on the skill and experience of him who used them.
Past studies of the GFRI in such diverse settings as the U.S.A.,
Canada, and Israel, had proved the reasonableness and usefulness of the tests.

Next it was shown that the results reflected nothing but data
volunteered by the group itself. Any weakness they revealed
were thus not imputed by outside critics, but disclosed in a
process of the group's self-examination. Such weaknesses our
interpretation showed to boil down to: (a) poor social cohesion; (b) a discrepancy between the community's cooperative
potential and the actual cooperativeness displayed in the community; (c) an incongruity between the subjective happiness
ratings and the unsatisfied communitarian needs. Together,
these weaknesses indicated a potentially dangerous deflection
from the goals which *Clermont* had set itself as a *community
of work*. The test showed that, instead of extending to both
work and personal relations, their cooperation tended to be
restricted to work only. For example, by far the largest number of all the mutual choices were made on work alone. Comparison of the sociometric with the cooperative potential
scores showed that the group in its present state tended to assign high status to members whose cooperative potential was
only fair or even low, and to relegate to the status of "neglectee" those whose cooperative potential was high. There was,
to be sure, nothing wrong with high regard for economic proficiency. If *Clermont* aimed at nothing more than a producers'
cooperative, the situation was acceptable as it was. However,
the task it had set itself was quite different and more difficult.
It had undertaken to demonstrate that industrial workers not
only could work but also live together cooperatively. The at-
tractiveness of the formula *Clermont* had evolved was attested
to by the fact that so many other groups had begun to emulate
it. A movement was set afoot that might become an important phase in the development of industrial civilization. The

success or failure of *Clermont,* which had started and was leading this movement, had consequences beyond its own fate. That failure as a cooperative *community of work* was not merely a remote possibility, was indicated in a particularly acute way by the test results regarding the family companions. As a subgroup, the tests indicated, they had developed as yet neither a social structure of their own nor any mutual ties with the productive companions. Such mutual ties as they did form among themselves were, moreover, related exclusively to work. This meant that the wives of the companions were little integrated in the community. Since, as the cooperative potential test showed, their score for all purposes equalled that of the productive companions, it could not be contended that their failure was simply a lack of cooperativeness. It had to be viewed rather as a failure on the part of the community to offer the family companions a reasonable chance to realize their cooperative potential. The same was true of the productive companions with the highest cooperative scores. It is only that with respect to the family companions this failure was more complete and therefore more alarming. They ought to consider that for the family companions, who take no direct part in economic production, the main, if not only, possibility for active participation in the community was in the neighborhood group. That group had been defined by the founders of *Clermont* as the elemental cell of the community. The explanation of the total absence of mutuality among the family companions themselves, and between them and the productive companions, lay in the fact that the neighbor groups were failing to perform the vital function for which they were set up. That, in turn, meant that the very roots of community were in jeopardy at *Clermont.*

If the first two tests came at all close to revealing the real situation, there could be no doubt that the danger was acute. The obstacles test indicated, however, that there was also the strength sufficient to meet it effectively. We could infer that the concern with lack of community spirit was well matched by the amount of resoluteness, or strength of will, to overcome it. It was hoped that the present study would help activate that will in the form of effective remedial action.

Steps of Remedial Action

The effect of the presentation was unmistakable. The most dramatic confirmation of its impact came when the leader of the clique which had agitated against the tests, rose in the Assembly and publicly expressed his regret for having obstructed the progress of the study, and thus possibly impaired the validity of its results. That admission, although the speaker himself probably did not realize it, was actually the first step in the direction of remedial action. Thereafter, the meeting turned into what it was intended to be, a group consultation concerned with the proper course of action. The points made in the presentation had indicated the steps to be taken. Speaking in the name of the community, its Head said the results of the tests substantiated what many of the members had vaguely felt. This was true particularly with regard to the neighborhood groups. Their revitalization undoubtedly was the most urgent task, to be undertaken immediately on the basis and with the help of the test results. Steps to raise the cooperative spirit of the community, the Head of the community revealed, already had been taken. In processing new applicants, the community had begun to employ the cooperative potential test in the sense that, of several candidates more or less equally qualified for a job, those with the higher cooperative scores were being accepted. In the ensuing discussion, further steps were proposed, among them mainly community arrangements, such as a community laundry, expansion of the nursery, and the like, that would help free the women from their household duties and allow them more time for active participation in community affairs.

Thus, without any further prodding, the group at once began

to plan its own therapy. By doing so, it chose what probably is the only effective way to redress a group's ailments: the group's own deliberate remedial action. By stirring the group up to it, the tests had served their diagnostic purpose.

14

SUMMARY AND OUTLOOK

Recent Developments

Three agencies devoted to research in the sociology of cooperation were founded during the summer of 1953. They are the *Bureau d'Etudes Coopérative et Communautaires,* Paris, France, the *Sektion fuer die Soziologie des Genossenschaftswesens* (Section for the Sociology of Cooperation) of the *Institut fuer Selbsthilfe,* Cologne, Germany, and the *International Council for Research in the Sociology of Cooperation,* Geneva, Switzerland. The activities of these new agencies may mark a significant advance in the field with which the Group Farming Research Institute is concerned. Before describing them it might be well, therefore, to summarize briefly what has been indicated previously about the nature and scope of the sociology of cooperation.

The Sociology of Cooperation

As the term suggests, sociology of cooperation implies the application of sociological methods to the study of cooperation. Simple though this may sound, unfortunately its meaning cannot be taken for granted, mainly because of the vagaries of use and abuse to which the terms "sociology" and "cooperation" still are subject. As to sociology, it has not yet quite overcome the state caustically referred to by Henry Poincaré in his *Science and Method* (1908). "Nearly every sociological thesis," he said, "proposes a new method, which, however, its author is very careful not to apply, so that sociology is the science with the greatest number of methods and the least results." Nevertheless, within the last generation, sociology, especially in America, has undergone a change for the better both in theory and practice. A growing number of sociologists

311

has come to accept the dynamics of human interaction as their specific common subject matter. Unlike the former more sedentary students of society, modern sociologists go into the field and search for Poincaré's "regular facts." One important result of this change in approach is a growing awareness of the group aspect of society and of closer attention to more precise methods and techniques of investigation. A new kind of sociology is developing today which, whether it calls itself "applied sociology," "sociometry," "clinical," or "microsociology," seeks to base itself on reasoning similar to that which has led to the astounding developments in nuclear physics. If it is true that the study of the smallest units produce the most precise insight into the nature of the universe, it appears likewise most promising to examine the smallest coherent units, or small groups, in order to understand society as a whole.

The student of cooperation accepts this position not only because it is scientifically sound but because it also accords so well with his specific field of inquiry, cooperation, viewed by the sociologist as a particular form of human interaction. He leaves it to biologists and to anthropologists to study cooperation as it manifests itself on the animal level or among "primitive peoples," and concerns himself chiefly with cooperation as instituted through the so-called Rochdale Principles and as advocated and practiced today by the cooperative movement. Sociological analysis of this practice shows that it can be of two basic kinds, mainly segmental or mainly comprehensive. In the first case, it means merely partial, and in the latter full or all-inclusive personal involvement in a cooperative association, as illustrated by membership in a consumers' cooperative store and in a *kvutza* respectively. Historically, the cooperative movement may be said to have passed through three different stages: the first, from the early groping beginnings at the end of the 18th century to about 1834; the second, from the opening of the consumers' store in Rochdale in 1844 and the formulation of effective principles for cooperative associations to the end of the first decade of this century; and the third and current stage, beginning with the establishment of the first modern cooperative community, the *Kvutza Dagania Aleph* in Israel, in 1911. Each of these stages receives its peculiar "style" from the kind of cooperative practice it favors. Thus the first stage may be seen as characterized by the exploration of the possibilities of both comprehensive (Robert Owen) and segmental cooperation (William King), leading to no lasting results in either case. The second stage was marked by the definite predominance and most effective extension of segmental cooperation (John T. W. Mitchell). In the third and present stage, comprehensive practice is gaining slowly but steadily a legitimate place side by side with the very solidly established practice of segmental cooperation.

Although the cooperative movement has spread to virtually all parts of the civilized world and by the beginning of World War II had attracted a membership of some 145,-000,000 people in 57 different countries, its influence is still felt very little in world affairs. The discrepancy between numerical strength and lack of persuasive vigor suggests some basic weakness in the nature or structure of the movement. Pertinent studies show that in the overwhelming majority of cooperatives the membership is motivated by purely economic considerations. The social aspects of cooperation are almost totally neglected. Attendance at general meetings is very poor and rarely exceeds two per cent of the membership. Thus failing to activate its own membership in anything beyond purely economic pursuits, it is hardly surprising that the cooperative movement fails to exert any influence on those who, for one reason or another, are not interested in the material advantages it has to offer.

The weakness in appeal of the cooperative movement may, indeed, be traced to its inertia in all matters relating to active participation by the members. The crucial question for those aiming to make cooperative practice, if not general, then at least as extensive as possible, is precisely the question of how to induce people to cooperate. Here social science may be of assistance. Before we try to answer the question of *how* to make people cooperate, we must inquire into *what* makes them do so. The discipline, generally concerned with human interaction, is sociology. However, because the specific form of interaction here called for is that of doing things together, the specific need is for a sociology of cooperation.

The systematic cultivation of this field of sociology has been the particular concern of the GFRI and this author. We have proceeded simultaneously on two levels, one purely fact-finding and theoretical, the other applied and "clinical." Advance on both levels was based on materials gained from direct contact with and study of cooperative communities, where-ever their development assumed significance, whether in the U.S.A., Canada, Mexico, Israel, or France. We have focussed our attention on the cooperative communities not because of any disregard for other types of cooperative practice, that is, those of a more segmental nature, but because of particular theoretical and practical considerations. Experience has shown that under conditions of comprehensive cooperation the "cooperative effect" was more pronounced than under the conditions of segmental cooperation and therefore could be observed and studied with greater facility and precision. An understanding of the workings of cooperative communities is of great practical importance wherever, as in many parts of the world today, the problem of increased food production is attacked by means of cooperative farming, a form of pro-ductive activity that cannot succeed well without some form of communal living. It turned out that the study of cooperative communities has enhanced our understanding of other types of cooperatives as well.

To facilitate, and at the same time to make more precise the observation of the cooperative effect, a number of research tools was developed and combined into a battery used today in all our studies. If it is true that the nature and scope of a discipline is most concretely circumsribed by the methods and techniques it employs, it might be said that the battery which we described in preceding pages of this volume, marks the boundaries reached today by the sociology of cooperation. Advance beyond them is both desirable and necessary. One way of achieving it is to reach out into new areas of investigation; another, to refine the existing methods and to add pertinent new techniques of inquiry. There is good reason to believe that the several newly established research centers will appreciably contribute to these objectives.

Bureau d'Études Coopératives et Communautaires (B.E.C.C.)

Following a study of eight French *communities of work* made by the GFRI in 1951,[1] one of the first tasks the new Bureau addressed itself to was that of improving the French wording of the tests and of adapting them more accurately to the given conditions, a task carried out in lengthy meetings which this writer attended during a trip to Europe in 1953.

By that time the Bureau had been in existence for about two months. It had been established following a resolution by the executive committee of the *Entente Communautaire* at a session in March 1953. The resolution of the *Entente* had recognized the need for research in the sociology of cooperation and declared itself ready to launch it by securing the necessary facilities for it. The executive committee was motivated by a realization of the importance of scientific analysis in the progress of the communitarian experiments and, indirectly, of the cooperative movement as a whole, and by educational considerations as well. To quote from the programmatic statement published under the heading "Un bureau

[1] Cf. "The Urban Cooperative Community" and "Experimental Groups and Sociological Counselling" in this volume.

d'études etc." in *Communauté* (No 1, 1953), by M. Henri Desroche, the moving spirit of the new venture:

> It might be taken as a good omen for the services the new Center will be able to render to science in general that it is being founded at a time when the need for a closer contact between research and practice finds official recognition in the recent appeal by our Minister of National Education. As far as the social sciences in particular are concerned, there hardly can be any doubt that only through such contact can there be hope that they may improve their own instruments of observation and analysis. In addition, there is the possibility of sociological experiment, for which the communitarian and similar groups are especially well suited. The paraphernalia of academic research can be of little use here. All they apparently are able to produce are actuarial arrays and platitudinous generalizations, and doctors theses that more often than not are limited to either. What is needed is a kind of research which, although as objective and precise as possible, requires insight into the inner workings of a group. In this way, inquiry is turned into a group action which not only helps the group to understand itself better and to pursue its proper goals with greater self-assurance, but which at the same time offers social science an opportunity of refining and perfecting its methods and techniques. Awareness of this opportunity induced some of the French, English, Swiss and American sociologists, especially those interested in such fields as the history and sociology of cooperation, the sociology of knowledge, the sociology of utopianism, and so on, to pay increased attention to groups which go by the name of 'cooperative community experiments.' Stimulated by this attention, the groups, on their part, are beginning to discover the advantages of scientific observation of their experimen-

tal undertakings and to welcome the assistance they may derive from continuous contact with a research body devoted to the study of issues which must be crucial to themselves.

The principal concerns which M. Desroches assigns to the new research center are: (1) the study of communitarian theory and practice of the past; (2) the study of present economic aspects and living standards; (3) the study of interpersonal and intergroup relations. Under the first heading would come such studies as the history of "utopian" socialism, of the cooperative movement with special attention to the role of comprehensive cooperation, and of worker's movements and socialism in general as they affect communitarian experiments. The studies planned under the second heading are to draw on source material concerning management, methods of remuneration, industrial psychology and sociology, ventures in profit sharing, and the like. Special attention is to be paid to the effect of communitarian organization on standards of living, diet and housing, family budgets, and last but not least, the status of women. For the studies of interpersonal and intergroup relations, it is stated that "use will be made of the findings of social psychology and the methods and techniques exemplified by the study of several *communities of work* conducted by the GFRI."

The range of activities may be too extensive and may surpass the capacities of the Bureau in its present stage. However, as M. Desroches puts it, "the attempt is in itself an experiment worth undertaking."

Other Developments

The establishment of the B.E.C.C. was made possible by the existence of an active communitarian movement in France. A recent trip to Europe revealed, however, that the need for the kind of research described above is felt in other countries as well. Interest was found to be keen in Italy, where possibilities were discussed with Dr. Adriano Olivetti, the leader of the *Movimento Communità,* though without leading to immediate action. In Germany, however, a *Sektion fuer die Soziologie des Genossenschaftswesens* was established in July 1953 at the *Institut fuer Selbsthilfe* in Cologne, thanks mainly to the initiative of its director, Professor Gerhard Weisser. Similar in purpose to the French Bureau, the *Sektion,* preliminary to the studies to be undertaken, already has prepared a translation of the battery into German. The first item on the agenda of the *Sektion* is a thorough study of the village of Huetschenhausen, which is probably the first instance of a "cooperative village" in post-war Germany.[2]

The study is planned as a joint enterprise of the *Sektion* and the Sociological Institute of the University of Cologne. The director of the latter, Professor René Koenig, intends to make the application of the battery in the study of the cooperative village part of a larger investigation concerned with the problem of whether, how and to what extent research methods developed in the U.S.A. are applicable to other countries, especially to Switzerland and Germany.

The interest in the sociology of cooperation which led to the establishment of two new research centers was surely gratifying in itself. It was felt to be desirable, however, that some organization be established to coordinate the activities of the several research centers. Hence, an initiating committee of interested parties met at Geneva and on May 30, 1953 incorporated the *International Council for Research in the Sociology of Cooperation,* under Swiss law. The purpose of the Council was stated briefly as follows: (1) to promote research in the sociology of cooperation, (2) to stimulate exchange of findings and services among organizations and students working in this field in different countries, (3) to

---

[2] Cf. This writer's "A German Cooperative Village," *Cooperative Living,* V, 2 (Winter 1953/54).

foster the dissemination and publication of research findings in this field, including translation into different languages.

Membership in the Council was declared open to scientific bodies and individuals interested in this field of research. At present the membership is composed of the three active research centers, the GFRI (U.S.A.), the B.E.C.C. (France), and the *Sektion* of the *Institut fuer Selbsthilfe* (Germany). A number of eminent social scientists in Britain, France, Germany, Israel, Italy, Switzerland, India and the U.S.A., have accepted membership in the Council and act as editors of the *International Library of the Sociology of Cooperation*, of which this book forms the first volume.

ACKNOWLEDGEMENTS

The articles and papers, on which these essays are based, appeared, for the largest part, in the past five volumes of *Cooperative Living,* the Bulletin of the Group Farming Research Institute and in: *Cooperative Group Living,* an International Symposium on Group Farming and the Sociology of Cooperation, Henry Koosis & Co., N. Y.; the *Journal of Human Relations,* issued by the Central State College, Wilberforce, Ohio; *The Year-Book of Agricultural Cooperation,* published by the Horace Plunkett Foundation, London, England; and the *Review of International Cooperation,* issued by the International Cooperative Alliance. Several of the articles appeared also in foreign-language publications, as in France: *Communauté,* the monthly of the Entente Communautaire; *Le Diagnostique Economique et Social,* of Economie et Humanisme; and *Coopération,* the monthly of the French Consumers Cooperatives; in *Israel: Kama,* the Year Book of the Keren Kayemeth L'Israel; and *Germany: Koelner Zeitschrift fuer Soziologie,* of the Sociological Institute of Cologne University, and *Archiv fuer Oeffentliches und Freigemeinwirtschaftliche Unternehmen,* the organ of several German organizations, issued in Cologne. Although the material has been considerably revised and completely rewritten in parts, it is a pleasant duty to give due credit to the publications in which it originally appeared.

For the sound advice, care, and painstaking attention paid to often tedious detail in readying the manuscript for publication, special thanks are due to my friend and co-editor of this series, Dr. Joseph Maier of Rutgers University.